The Tangled Bank: An Introduction to Evolution

Second Edition

Study Guide

Alison E. H. Perkins

Roberts and Company
Greenwood Village, Colorado

The Tangled Bank: An Introduction to Evolution, Second Edition, Study Guide

Roberts and Company Publishers, Inc.
4950 South Yosemite Street, F2 #197
Greenwood Village, CO 80111 USA
Tel: (303) 221-3325
Fax: (303) 221-3326
Email: info@roberts-publishers.com

Internet: www.roberts-publishers.com

Publisher: Ben Roberts
Production director: Julianna Scott Fein
Creative director: Emiko-Rose Paul
Artist: Tom Webster at Lineworks
Cover designer: Emiko-Rose Paul
Proofreader: Dawn Hall

Front cover: *Ambulocetus*, by Carl Buell
Credits: p. 16, Shutterstock.com/Eric Issele; p. 151, *Ethology 108*, "Environmental Predictors of Geographic Variation in Human Mating Preferences," Abstract, Kevin J. McGraw. Copyright © 2002 Blackwell Verlag, Berlin. Reproduced with permission of Blackwell Publishing Ltd.; p. 156, left: Shutterstock.com/Arto Hakola; center: All Canada Photos/Superstock, Inc./Tim Zurowski; right: Shutterstock.com/Steve Byland; p. 175, iStockphoto.com/gvictoria; p. 176, iStockphoto.com/Steve Byland; p. 192, H. Douglas Pratt.

ISBN: 9781936221455

==
10 9 8 7 6 5 4 3 2

CONTENTS

INTRODUCTION TO THE TANGLED BANK STUDY GUIDE

This study guide is a tool to help you understand what you've learned reading *Tangled Bank: An Introduction to Evolution*, by Carl Zimmer. It's designed to be engaging, using games and exercises and online resources to broaden your experiences. You can dive in as deeply as you choose, doing some exercises and skipping others, and you can test your comprehension when you're done.

Check Your Understanding

The *Check Your Understanding* questions are designed to remind you of important concepts—concepts that the current chapter builds on. Use the questions and answers at the end as a review, and if you struggle too much, pop back to previous chapters and revisit the concepts.

Learning Objectives

➢ Learning objectives allow you to identify important ideas presented in the text.
➢ Learning objectives prompt you to reach certain goals with your thinking.
➢ Learning objectives help identify the skills you should develop as a learner.

Identify Key Terms

The chapters in *Tangled Bank* introduce and discuss important concepts related to the theory of evolution. Here, you are challenged to match key terms with their definitions. These key terms are also important to developing concept maps and linking concepts in the next section.

Link Concepts

Each chapter includes a tool, either a concept map or another visual activity, to help you organize part of the chapter in your mind. Concept maps can be difficult to "get," but really, there is no "right" answer—concept maps are works in progress. The idea is to help you visually link ideas outlined in the chapter, and gradually build an understanding of how those ideas go together. The arrows used to link the concepts don't necessarily mean that one concept "leads to" another—simply that one concept *is related to* another in some way. You may prefer to think of the links in the opposite direction, or maybe you have a different way of linking the concepts altogether. The important thing is to think about how concepts you've learned are linked and *why*, instead of just jumbling everything up. Talking about concept maps with classmates or friends can help you develop your ideas and improve your understanding.

The first concept map is outlined, showing one way of thinking about how the relationships are delineated, and the key concepts that need to be considered are listed to the side. To take it to the next step, you are often challenged to come up with your own concept map for another part of the chapter. Concept maps can be great tools for visual learners!

Interpret the Data

Scientists rely on visual representations of data to help others understand their evidence—to tell their stories. But charts and diagrams can be difficult to grasp, and interpreting graphs takes practice. It's often too easy to read the caption and skip over actually evaluating the data presented. Check to make sure you understand the graph from the textbook presented in this section. You'll gain insight into the kinds of evidence scientists gather and how scientists use that evidence.

Games and Exercises

The study guide includes some hopefully fun, but definitely explanatory, games that demonstrate the principles important to understanding evolution. Some can be done alone, and some require the help of friends or classmates. The games often involve jelly beans, M&Ms, or money, so don't eat or spend until the end of the book!

Go Online

The Internet is an incredible resource for videos, data, analyses—and baloney—about evolutionary biology. Some of the resources included in the study guide offer stunning visuals (PBS *Evolution*, *NOVA*, National Geographic), and even YouTube videos (although beware—YouTube and the Internet are full of misinformation, too). For cutting edge scientific studies, check out the links to *Science* magazine from the American Association for the Advancement of Science (AAAS). The links to *Science*Now are recent news stories that provide readable summaries and explanations of the current state of scientific research. Check the AAAS website regularly to find even more recent examples of the work scientists are doing.

The QR Codes
QR (short for "quick response") codes are essentially bar codes that can be read by imaging devices, such as cameras. If your smartphone has a camera, there's likely an app available that you can use to read the QR codes.

The QR codes used in the study guide use the URLs to take you to the websites listed in the text. The codes were generated using GoQR.ME, created by Andreas Haerter and Andreas Wolf and should be completely free of advertising. "Like" them!

Common Misconceptions

The theory of evolution involves some abstract and complex concepts, and these concepts can be particularly difficult to grasp. Some misconceptions result simply because the ideas were never explained very well, and others because they seem to make sense—even though they are misconceptions. Other misconceptions are actively played up in the media, on the Internet, and by sources motivated to discredit science. Think carefully about the common misconceptions outlined in each chapter, and make sure you understand why they are misconceptions.

Contemplate

Sometimes it's important to just think about stuff, and the questions posed in this section give you the opportunity to get a little creative with your thinking. Some questions really don't have answers, but share your ideas with classmates and friends. You might be surprised at what you can come up with.

Delve Deeper

You can really test your understanding with these short answer questions. They are designed to help you put concepts together and to address some of the confusion that may arise as you learn more and more about the theory of evolution. See how many you can tackle successfully.

Go the Distance: Examine the Primary Literature

Scientists rely on the primarily literature for communicating their results. Thousands and thousands of research papers are published every year, and sifting through all that evidence can feel overwhelming. Check out the research paper highlighted for each textbook chapter, and see if you can answer the questions.

Test Yourself

Multiple choice questions at the end of each study guide chapter are a great quick assessment of your understanding of the basic ideas presented in the text.

Answers

Of course answers to almost everything can be found at the end of each study guide chapter. The Check Your Understanding questions go a little further and help you out if you're unclear about the right answer and why. If you are still uncomfortable with the concepts, you should probably brush up before moving on.

There are no right answers for the concept maps, so don't expect to find an answer to the challenge to develop your own map. Nor will you find answers to questions posed in the Contemplate section. You should talk to your friends and classmates about their ideas on these two sections—you'd be surprised how informative those conversations can be.

Similarly, the Games and Exercises may or may not give you answers consistently. That's the beauty of data and evidence and interpretation. Many experiments need to be conducted over and over and over until a trend emerges. Sometimes random processes are more influential than experimental design, and sometimes experimental error affects results. Talk about the results you get with your classmates or your friends—compare notes, procedures, ideas—and then think about what it means to do science.

ACKNOWLEDGEMENTS

Carl Zimmer is an amazing writer, and I'm so glad to have worked with him on this study guide. I hope this study guide will complement his incredible contribution and add to readers' understanding of evolution. Doug Emlen also was instrumental in the development and review of the study guide, and I cannot thank him enough for taking the time to provide thoughtful insight and editing. Several other reviewers provided valuable assistance crafting the study guide, including Cerisse Allen, Marc Albrecht, and Joel McGlothlin.

I am also incredibly grateful to Emiko Paul for all her help and support, Julianna Scott Fein who took the reins and made things happen, Tom Webster who helped with the art, and, of course, to Ben Roberts.

Most importantly, thanks to all the evolution and science educators who are developing curricula and websites to help people understand the complexities of evolutionary theory and the scientific endeavor. So much incredible material is available, and this guide would not be possible without contributions from the American Association for the Advancement of Science (AAAS), Janis Antonovics and Doug Taylor of the Biology Department at University of Virginia, Ball State University Electronic Field Trips, Cornell Lab of Ornithology, Darren Fix (ScienceFix.com), Larry Flammer and the Evolution and the Nature of Science Institutes (ENSI), Judy Parrish, the Smithsonian, Society for the Study of Evolution, and Tracy Tomm (ScienceSpot.net).

REFERENCES

Brennan, P. L. R., R. O. Prum, K. G. McCracken, M. D. Sorenson, R. E. Wilson, et al. 2007. Coevolution of male and female genital morphology in waterfowl. *PLoS ONE* 2 (5):e418-e418.

Burger, J. M. S., M. Kolss, J. Pont, and T. J. Kawecki. 2008. Learning ability and longevity: A symmetrical evolutionary trade-off in *Drosophila. Evolution* 62 (6):1294-1304. doi: 10.1111/j.1558-5646.2008.00376.x.

Coyne, J. A. and H. A. Orr. 2004. *Speciation*. Sinauer Associates, Sunderland, MA. 545 pp.

Ding, L., T. J. Ley, D. E. Larson, C. A. Miller, D. C. Koboldt, et al. 2012. Clonal evolution in relapsed acute myeloid leukaemia revealed by whole-genome sequencing. *Nature* 481 (7382):506-510. doi: 10.1038/nature10738.

Flynn, J. J., J. A. Finarelli, S. Zehr, J. Hsu, and M. A. Nedbal. 2005. Molecular phylogeny of the Carnivora (Mammalia): Assessing the impact of increased sampling on resolving enigmatic relationships. *Systematic Biology* 54 (2): 317–37.

Gatesy, J. and M. O'Leary. 2001. Deciphering whale origins with molecules and fossils. *Trends in Ecology and Evolution* 16: 562-570

Gregory, T. 2008. Understanding evolutionary trees. *Evolution: Education and Outreach* 1:121-137.

Herrmann, E., C. Josep, H.-L. Maráa Victoria, H. Brian, and T. Michael. 2007. Humans have evolved specialized skills of social cognition: The Cultural Intelligence Hypothesis. *Science* 317 (5843):1360-1366. doi: 10.1126/science.1146282.

Losos, J. B., and R. E. Ricklefs. 2009. Adaptation and diversification on islands. *Nature* 457:830-836.

Meir, E., J. Perry, J. C. Herron, and J. Kingsolver. 2007. College students' misconceptions about evolutionary trees. *The American Biology Teacher* 69(7):e71-e76.

Tishkoff, S. A., F. A. Reed, F. R. Friedlaender, C. Ehret, A. Ranciaro, et al. 2009. The genetic structure and history of Africans and African Americans. *Science* 324 (5930):1035-1044. doi: 10.1126/science.1172257.

Chapter 1
WALKING WHALES: INTRODUCING EVOLUTION

Check Your Understanding

1. Because evolution is a theory, that means:
   a. It is a guess or a hunch.
   b. It has very little evidence to support it.
   c. Scientists may or may not believe in it.
   d. It will explain God.
   e. None of the above.

2. Why is understanding evolution important?
   a. Understanding evolution can help us understand biodiversity issues associated with deforestation and global warming.
   b. Understanding evolution can help us understand the evolution of drug resistance and cancer.
   c. Understanding evolution can help us understand our own DNA and how it affects our lives.
   d. All of the above.

Learning Objectives for Chapter 1:

> ➤ Define evolution.
> ➤ Describe two basic principles that make evolution possible.
> ➤ Explain the relationship between artiodactyls and modern whales.

Identify Key Terms:

Connect the following terms with their definitions on the right:

adaptation	Descent with modification.
artiodactyl	An explanation that is proposed for an observation, phenomenon, or scientific problem that can be tested with further examination.
cetacean	A member of the order Cetacea—mammals including whales, porpoises, and dolphins.
evolution	A group of mammals (including cows, goats, camels, and hippos) whose astragalus has a double-pulley structure and that bear their weight equally on the third and fourth toes.
hypothesis	A change to the structure or function of a part of an organism that resulted from natural selection.
involucrum	The inner wall of the ectotympanic; an important characteristic in the evolution of whales.

Link Concepts

1. Fill in the following concept map with the key terms from the chapter:

evidence

fossils

scientific test

DNA

hypothesis

provides   anatomical

genetic

generates

leads to

predicts

Cetacean Evolution

2. Develop your own concept map that explains the relationship between artiodactyls and modern whales.

Interpret the Data

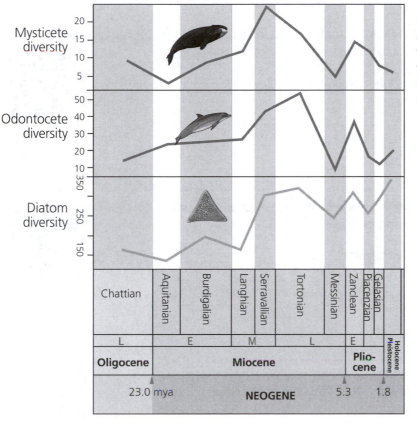

Figure 1.10B shows the increases and decreases in the diversity of whales over the past 55 million years. Systems are listed on the bottom line, series on the second line, and stages above those.

When did the diversity of baleen whales (mysticetes) peak?

When did the diversity of toothed whales (odontocetes) peak?

When was the diversity of diatoms highest?

From this graph, what might you predict about the diversity of dolphins (odontocetes) in the future? What is your prediction based on?

Can you predict the same about the diversity of whales (mysticetes)?

What factors might affect the accuracy of your predictions?

The Tangled Bank Study Guide

Games and Exercises

Evolution in Action

An interactive game from PBS NOVA that allows you to see how random mutations can lead to the evolution of a population of virtual creatures in different environments.

http://www.pbs.org/wgbh/nova/evolution/evolution-action.html

Whale Kiosk

An interactive site that can help you understand the evidence for whale evolution created by Lara Sox-Harris.

http://www.indiana.edu/~ensiweb/lessons/whalekiosk.html

Guess the Embryo

This interactive game from PBS *NOVA* uses images of different embryos and the development of species to help show that embryos of different species can appear startlingly similar to one another, and the embryos of many species may or may not resemble their adult forms.

http://www.pbs.org/wgbh/nova/evolution/guess-embryo.html

Go Online

Return to the Water: Inside the Blue Whale

From the *Life of Mammals—Return to the Water* episode, produced by the BBC, this video takes you inside the blue whale, exploring the anatomy of the blue whale and its adaptations to life in the water.

http://www.bbc.co.uk/programmes/p004t035

Evolution: Great Transformations

This video from the PBS Evolution Library and WGBH shares the fossil evidence that supports the evolutionary transformation of whales and the characteristics that align whales with mammals not fish or sharks.

http://www.pbs.org/wgbh/evolution/library/03/4/l_034_05.html

Fellowship of the Whales: Cooperative Feeding in Humpback Whales

PBS Nature provides fantastic footage of cooperative hunting in humpback whales.

http://www.pbs.org/wnet/nature/episodes/fellowship-of-the-whales/video-cooperative-feeding/5324/

You can watch the full episode of the *Fellowship of the Whales* here:

http://www.pbs.org/wnet/nature/episodes/fellowship-of-the-whales/introduction/5263/

From National Geographic Kids, video from a camera placed directly on the blue whales as they hunt.

http://video.nationalgeographic.com/video/kids/animals-pets-kids/mammals-kids/whale-humpback-kids/

Dolphin Language: Two Dolphins Communicate with Each Other

This YouTube video shows trainers working with two captive dolphins as they communicate and learn a new trick.

http://www.youtube.com/watch?v=YSjqEopnC9w

New Whale Species Unearthed in California Highway Dig

Science Now reports the discovery of four new fossil species that offer new ideas about when toothed baleen whales went extinct.

http://news.sciencemag.org/sciencenow/2013/02/new-whale-species-unearthed-in-c.html

Common Misconceptions

Outside of science, evolution is often characterized as something it is not. Descent with modification can, and has, produced an amazing diversity of life, but this simple process does not necessarily fit the way we want to look at nature.

Peaceful Balance
Evolution is not about creating a peaceful balance among organisms. Organisms eat other organisms, and that may sometimes be gruesome in our eyes. The same processes that give rise to exquisite adaptations, though, also give rise to the amazing capabilities of predators for finding and killing prey. Parasites may also have amazing adaptations, such as the capacity to manipulate individual molecules within their hosts that allow them to devour their hosts from the inside out. Predation and parasitism may seem cruel, but

predators and parasites are not evil. These organisms are simply responding to the same selective pressures that all organisms face, driving both the diversification and extinction of life.

Adaptations Species "Need"

The match between species and their environments can be striking, but this match did not come about because the species needed specific adaptations. Evolution cannot determine need—after all, one organism may "need" to live and breed, but another organism may "need" to kill and eat the first organism. Evolution is based on simple processes, such as random mutations that become more or less common over the course of generations. An adaptation in one individual that performs

better than in another individual should result in more offspring that have the genetic architecture that produced the better version of the adaptation in the next generations—better and better adaptations become more and more common.

Perfect Adaptations

As impressive as some adaptations may be, they are far from perfect. Adaptations do not evolve from scratch; evolution modifies what already exists. Because beneficial mutations are limited, new forms evolve under tight constraints. Plus, mutations can have several different effects at once, so evolution also involves trade-offs. Many of the diseases we face, such as cancer, result from these trade-offs.

Contemplate

Did whales evolve *into* humans? What about apes—did they evolve *into* humans?	Are embryos miniature versions of the animals they eventually become?

Delve Deeper

1. How can mutated genes become more or less common over the course of generations?

2. Do you agree that evolution produces a peaceful balance in the natural world?

3. What evidence can scientists use to test hypotheses about how different species are related to each other?

4. What characters distinguish whales from sharks and tuna?

5. The text states, "Evolution is imperfect because it does not invent things from scratch: it only modifies what already exists." What evidence from whale evolution supports this statement?

Go the Distance: Examine the Primary Literature

Hans Thewissen and his colleagues examine how genes affect development of hind limbs in modern dolphins (yes, dolphin embryos begin to grow legs). They piece together a hypothesis about both the small changes (genes) and large changes (species) in their evolutionary history.

- What evidence did they use?

- How did they develop their hypothesis about whale evolution?

Thewissen, J. G. M., M. J. Cohn, L. S. Stevens, S. Bajpai, J. Heyning, and W. E. Horton Jr. 2006. Developmental Basis for Hind-Limb Loss in Dolphins and Origin of the Cetacean Body Plan. *Proceedings of the National Academy of Sciences* 103 (22): 8414–18. http://www.pnas.org/content/103/22/8414.full.

Test Yourself

1. Which of the following is a definition of evolution?
   a. Mutated genes becoming more or less common over the course of generations.
   b. Any change in the inherited traits of a population that occurs from one generation to the next.
   c. Descent with modification.
   d. All of the above are possible definitions.
   e. None of the above is an appropriate definition.

2. Which group of animals are whales most closely related to?
   a. Giraffes.
   b. Tuna.
   c. Sharks.
   d. Otters.
   e. None of the above.

3. Which is the best explanation for why mysticetes still have genes for building teeth?
   a. The genes now function to make baleen.
   b. They inherited these genes from ancestral whales.
   c. Mysticetes might need teeth, so evolution keeps the genes around.

   d. Evolution can't take away genes, only add new ones.
   e. All of the above are valid explanations.
   f. None of the above is accurate.

4. What characters of *Dorudon* are similar to modern whales?
   a. Presence of flippers.
   b. A long vertebral column.
   c. The shape of the involucrum.
   d. All of the above are characters shared by *Dorudon* and modern whales.
   e. None of the above is a character shared by *Dorudon* and modern whales.

5. Why was finding the fossils of *Pakicetus* so important in understanding whale evolution?
   a. Because the fossils made the researchers famous.
   b. Because *Pakicetus* shared many whale traits, but it lived on land.
   c. Because *Pakicetus* was not really a whale because it lived on land.
   d. Because the fossils weren't old enough to be considered the common ancestor of whale

Answers for Chapter 1

Check Your Understanding

1. Because evolution is a theory, that means:
   a. It is a guess or a hunch.
      Incorrect. Many people use the word *theory* in place of *guess* of *hunch*, but to scientists, the word theory carries a lot of weight. Evolutionary theory is well supported by evidence and years of examination, experimentation, and testing. *The Tangled Bank* will introduce you to some of that evidence.
   b. It has very little evidence to support it.
      Incorrect. There is much evidence to support the theory of evolution from Darwin's own research to the development of the modern synthesis, incorporating evidence from diverse fields such as geology, genetics, and geography, among others. *The Tangled Bank* will introduce you to some of that evidence.
   c. Scientists may or may not believe in it.
      Incorrect. Science is not about "believing" in theories; it moves forward based on the weight of evidence, and evolutionary theory is well supported by evidence and years of examination, experimentation, and testing. *The Tangled Bank* will introduce you to some of that evidence.
   d. It will explain God.
      Incorrect. Evolution does not even try to explain God; science focuses on explaining the natural world alone. Evolutionary theory is well supported by evidence from the natural world and years of examination, experimentation, and testing. *The Tangled Bank* will introduce you to the evidence, as well as the historical development of the theory. Understanding and accepting that evidence does not concern any explanation of God.
   e. None of the above.
      Correct. Because evolution is a theory, like Newton's theory of gravitation, it functions as a set of overarching mechanisms and principles that explain a major aspect of the natural world. Evolutionary theory is well supported by evidence and years of examination, experimentation, and testing; it explains the diversity of life on Earth. *The Tangled Bank* will introduce you to some of that evidence.

2. Why is understanding evolution important?
   a. Understanding evolution can help us understand biodiversity issues associated with deforestation and global warming.
      Correct, but so are other answers. The world's biodiversity faces many threats. *The Tangled Bank* will introduce you to how studying past mass extinctions can help us understand the effects our actions may have on the world's biodiversity.
   b. Understanding evolution can help us understand the evolution of drug resistance and cancer.
      Correct, but so are other answers. Through the process of evolution, some dangerous bacteria have become resistant to even the most powerful antibiotics. *The Tangled Bank* will explain how through observations of the evolution of resistance in laboratory experiments, scientists may be able to offer solutions to help slow resistance in nature.
   c. Understanding evolution can help us understand our own DNA and how it affects our lives.
      Correct, but so are other answers. Our species, and our DNA, is a product of the evolutionary process. *The Tangled Bank* will show how understanding evolution may provide answers to what it means to be human, how our DNA interacts with our environment, and what ancient segments lurk in our genomes.
   d. All of the above.
      Correct. Understanding evolution can help us understand biodiversity issues facing our society. It can help us understand resistance to antibiotics, and it can help us understand how our DNA makes us human. *The Tangled Bank* will introduce you to the theory of evolution, the evidence, experiments, and critical examination that have led to our current understanding, and the importance of understanding evolution to our future on this planet.

Identify Key Terms

adaptation	A change to the structure or function of a part of an organism that resulted from natural selection.
artiodactyl	A group of mammals (including cows, goats, camels, and hippos) whose astragalus has a double-pulley structure and that bear their weight equally on the third and fourth toes.
cetacean	A member of the order Cetacea—mammals including whales, porpoises, and dolphins.
evolution	Descent with modification.
hypothesis	An explanation that is proposed for an observation, phenomenon, or scientific problem that can be tested with further examination.
involucrum	The inner wall of the ectotympanic; an important characteristic in the evolution of whales.

Link Concepts

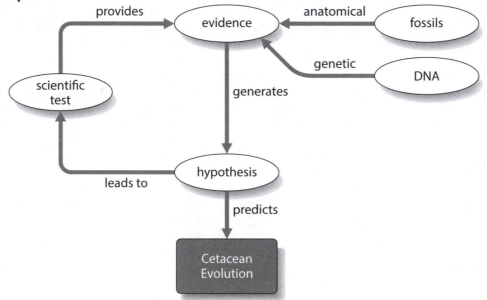

Interpret the Data

- When did the diversity of baleen whales (mysticetes) peak?
  During the Serravallian about 12 million years ago (see the systems line) and then again in the Zanclean around 4 million years ago.
- When did the diversity of toothed whales (odontocetes) peak?
  During the Tortonian around 10 million years ago and then again in the Zanclean around 4 million years ago.
- When was the diversity of diatoms highest?
  During the Tortonian around 10 million years ago, again in the Zanclean around 4 million years ago, and in the Holocene/Pleistocene (present stage).
- From this graph, what might you predict about the diversity of dolphins (odontocetes) in the future? What is your prediction based on?
  That the diversity of dolphins (odontocetes) will continue to increase. The prediction is based on the correlation between odontocetes and diatom diversity in the past and the current increase in both odontocetes and diatom diversity in the Holocene/Pleistocene.

- Can you predict the same about the diversity of whales (mysticetes)?
  You could, but the graph indicates a continuous decline in mysticetes diversity despite the increase in diatom diversity in the Holocene/Pleistocene (present stage). So even though mysticetes diversity may increase in the future, the evidence from fossil diversity does not support the prediction.

Delve Deeper

1. How can mutated genes become more or less common over the course of generations?
   A mutation may be lethal, it may be harmless, or it may be beneficial in some way. Mutations that are lethal should become less common over the course of generations because individuals with lethal mutations shouldn't do very well relative to other individuals. But if a beneficial mutation helps an organism to fight off diseases, to thrive in its environment, or to improve its ability to find mates, that individual should produce more offspring than individuals without the mutation.

2. Do you agree that evolution produces a peaceful balance in the natural world?
   No. Evolution does not produce a peaceful balance, nor does it strive to. Predators hunt and kill prey, and parasites devour their hosts from the inside out. The actions of predators and parasites may seem gruesome to us, but they are part of the natural process of evolution.

3. What evidence can scientists use to test hypotheses about how different species are related to each other?
   Scientists make comparisons between living species and fossils of extinct species to study shared anatomical traits. More recently, they began comparing DNA. Close relatives will share more traits, both physical and genetic, inherited from their common ancestor. So scientists can develop hypotheses for how these traits changed over time and how they evolved between different lineages and test them with additional evidence.

4. What characters distinguish whales from sharks and tuna?

Although whales have fishlike bodies, with the same sleek curves and tails you can find on tunas and sharks, they have a number of distinguishing characters. Whales do not have gills, so they cannot extract dissolved oxygen from the water in which they live. Whales must rise to the surface of the ocean in order to breathe. Whales and dolphins have long muscles that run the length of their bodies—much like the long muscles running down your back—whereas tuna have relatively simple sets of muscles that form vertical blocks from head to tail. Whales lift and lower their tails to generate thrust. Sharks and tunas move their tails from side to side. And whales give birth to live young that cannot get their own food; instead, the young must drink milk produced by their mothers. Only some species of sharks and fish give birth to live young, but those offspring can feed themselves; sharks and tuna do not produce milk. So despite their fishlike appearance, whales are very different from sharks and tuna.

5. The text states, "Evolution is imperfect because it does not invent things from scratch: it only modifies what already exists." What evidence from whale evolution supports this statement?
   The best evidence is from the blowhole that a whale breathes through. Rather than being able to obtain oxygen from water like a fish, a whale must breathe air like any other mammal. Since whales have to breathe air, they can only stay underwater for a limited amount of time. The nostril that its ancestors breathed through was modified into a blowhole. Other modifications include the flippers, the loss of hind legs, and the long muscles that run down the sides of their bodies.

Test Yourself

1. d; 2. a; 3. b; 4. d; 5. b

Chapter 2
BEFORE AND AFTER DARWIN: A BRIEF HISTORY OF EVOLUTIONARY BIOLOGY

Check Your Understanding

1. How can evolution be defined?
   a. Evolution is any change in the inherited traits of a population that occurs from one generation to the next.
   b. . Descent with modification.
   c. Mutated genes becoming more or less common over the course of generations.
   d. All of the above are possible definitions.
   e. None of the above is an appropriate definition.

2. What kinds of evidence *haven't* scientists used to study the evolution of whales?
   a. Evidence from DNA.
   b. Evidence from fossils, such as their anatomical traits.
   c. Evidence from geology.
   d. Evidence from living species.
   e. All of the above.

3. What trait would you consider the most important in determining whether a fossil should be considered a cetacean or not?
   a. The teeth because modern whales have peg-like teeth, but *Dorudon*, one of the earliest cetacean fossils completely adapted to life in the water, had diverse teeth.
   b. The involucrum because it has a unique form in cetaceans not found in other mammals.
   c. The astragalus because if cetaceans evolved from artiodactyls, then early cetaceans should have double-pulley astragali.
   d. The presence of flippers because the earliest whale fossils had flippers much like those of modern whales.
   e. The absence of legs because cetaceans do not have legs.

Learning Objectives for Chapter 2:

➤ Identify early naturalists and their contributions to evolutionary theory.

➤ Analyze the role the fossil record played in the development of the concept of evolution.

➤ Analyze how Darwin's observations of nature led to the inferences he developed regarding natural selection.

➤ Define theory as it is used in science.

➤ Identify some misconceptions about the theory of evolution.

Identify Key Terms

Connect the following terms with their definitions on the right:

artificial selection	The science of describing, naming, and classifying species of living or fossil organisms.
common descent	A mechanism that can lead to evolution because differences in individuals cause some of them to survive and reproduce more effectively than others.
extinction	The study of prehistoric life.
homologies	The permanent loss of a population or species, arising with the death or failure to breed of the last individual.
natural selection	The study of layering in rock (stratification) as a method for reconstructing the past.
paleontology	The theory that all species on Earth are related, like cousins in a family tree.
stratigraphy	An overarching set of mechanisms or principles that explain a major aspect of the natural world.
taxonomy	A mechanism that can lead to evolution because humans choose which individuals survive and reproduce because they have specific traits that are preferred for economic or aesthetic reasons.
theory	Traits that are similar because they are inherited from a common ancestor.

Color the homologous bones the same color in each of these forelimbs.

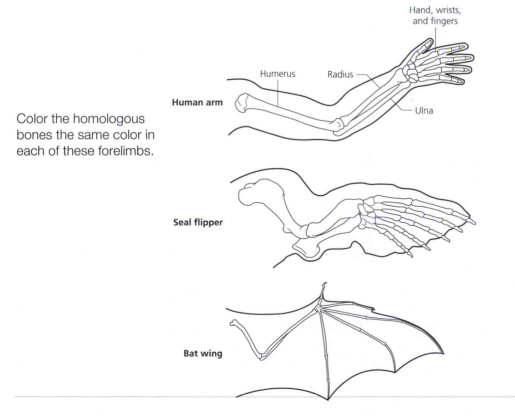

Hand, wrists, and fingers

Humerus   Radius

Human arm

Ulna

Seal flipper

Bat wing

Link Concepts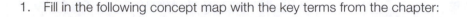

1. Fill in the following concept map with the key terms from the chapter:

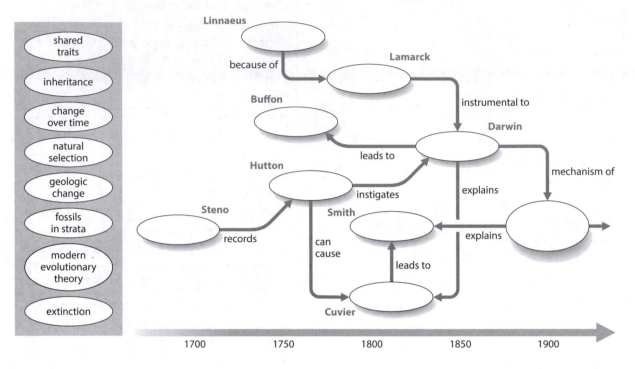

2. Develop your own concept map that explains the relationship between artiodactyls and modern whales.

Interpret the Data

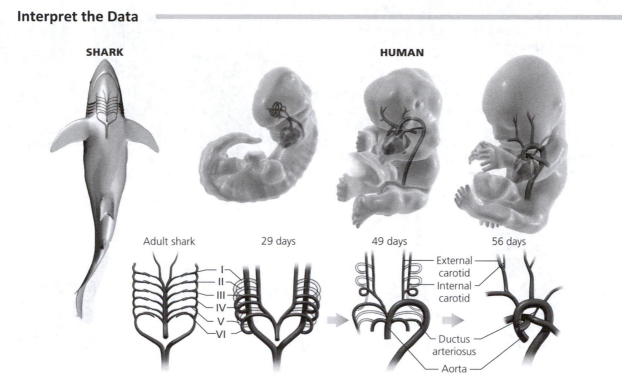

Figure 2.15 shows the arrangement of branching blood vessels used to absorb oxygen in adult sharks and human embryos at various stages of development.

At which stage of human development does the pattern of blood vessels look most similar to the pattern found in sharks?

If scientists were only to examine human embryos at 56 days, would the homologies be apparent?

What does this homology indicate about the relationship between fishes and humans?

Games and Exercises

How good a predator are you?
Survival and reproduction are not random. An individual with one version of a trait that helps it to thrive in its environment, such as better camouflage, may contribute more offspring to the next generation than other individuals—and those offspring are more likely to have that version of the trait. You can test how well certain individuals with certain traits survive in certain environments by pretending to be a predator to see how well you do spotting your prey. It's pretty easy to test you and your friends' abilities as

predators—you need individuals that vary in color and a background that serves as camouflage for some individuals and not others.

Go to the craft store and buy some red, white, and green pipe cleaners. Twist 2–3 pipe cleaners together using some pipe cleaners that are the same color and some of different colors until you have 20 "animals." Take the animals outside and place them in bushes, on tree branches, on the ground, and such, along a stretch of pathway.

 This activity is adapted from "Pipe Cleaner Animal Camouflage" by T. J. Fontaine from

http://www.bioed.org/ECOS/inquiries/inquiries/FCCamo.pdf.

Invite a couple friends to walk the stretch of pathway, at a moderately slow pace, and list the colors of animals they can find. Can they find all 20? If not, which colors were they more likely to find? Which were harder to find?

 You can also do a similar experiment with green and plain-colored toothpicks scattered in the grass in your yard in an exercise designed by Don Dunton, Fred Fisher, and Larry Flammer at ENSI (http://www.indiana.edu/~ensiweb/lessons/ns.st.wm.html). Spread equal numbers of green and plain toothpicks randomly in the grass.
Give you and your friends one minute to collect as many as you can, and then determine the proportion of green versus plain toothpicks you each located. Did the green trait affect the survival of some individual more than the plain trait?

 Alternatively, you can play *Nowhere to Hide*, an interactive online game that illustrates the same point.

http://sciencenetlinks.com/tools/nowhere-to-hide/.

From the Science Channel, journey with Charles Darwin as he explains natural selection and play an interactive selection game.

http://science.discovery.com/games-and-interactives/charles-darwin-game.htm.

Go Online

A Film about Carl Linnaeus

Learn more about Carl Linnaeus and his influence on modern-day taxonomy in this biography produced by the Natural History Museum in London.

http://www.youtube.com/watch?v=Gb_IO-SzLgk

Fossil Great White Shark from Antwerp

This short video shows a tongue stone, and the discovery of a fossil great white shark, Carcharodon carcharias, from the Pliocene, in Port of Atwerp, Belgium.

http://www.wat.tv/video/fossil-great-white-shark-from-4thu5_2htaj_.html

Time and Death—The Secrets of Evolution with Sagan, Cuvier, Darwin, Eiseley, and Barlow

Connie Barlow gives a really interesting lecture about the history of Carl Sagan's claim that "the secrets of evolution are time and death." The video begins with the insight of Georges Cuvier that species have indeed gone extinct in the past.

http://www.youtube.com/watch?v=mnTAzhLIEIg

Darwin's Dangerous Idea

From PBS, this short video introduces you to Charles Darwin and his voyage on the *Beagle*. Darwin uses the scientific process to amass evidence contrary to current ways of thinking, ultimately leading to publication of On the Origin of Species.

http://www.pbs.org/wgbh/evolution/darwin/index.html

Common Misconceptions

Opponents of evolution often make many misleading claims about evolution, BUT:

⇨ *Evolutionary Biologists Do Not Study How Life Began*
People who oppose evolution often state that "evolutionary biologists have not discovered how life began." Evolutionary biologists do not study how life began but how life diversified *after* it

began; other scientific fields of study are asking fruitful questions about how life began.

⇨ *Our Planet Is Not a Closed System*
People who oppose evolution often claim that "evolution violates the second law of thermodynamics." The second law of thermodynamics only holds in closed systems, but our planet receives outside energy from the sun—it is not a closed system.

⇨ *Evolution Is Not an Entirely Random Process*
People who oppose evolution state that "it is statistically impossible for evolution to have produced complex molecules like hemoglobin or organs like the eye," but that statement is based on the presumption that random processes led to a complex adaptation. Evolution is not an entirely random process; all the mutations that build an adaptation do not appear in a single event. Evolution occurs because mutations with beneficial effects spread through populations. Adaptations evolve as beneficial mutations accumulate over time.

⇨ *Mutations Can Easily Provide the Raw Material for Innovations*
People who oppose claim that "natural selection cannot produce innovations, because it simply favors some preexisting variants over others."

Mutations can easily provide the raw material for innovations—genes can be duplicated and mutations can lead to different responses or responses to different signals, any of which can produce new gene networks that can take on entirely new functions and even produce new structures.

⇨ *Gaps in the Fossil Record Should Be Expected*
People who oppose evolution argue that evolution fails as an explanation because of "absence of fossils that document major transitions." Paleontologists are actively exploring the fossil record, adding to our understanding of the relationships among taxa. But fossilization is not a perfect process—not everything that dies is preserved.

The Role of Debate in Science
Debate is a critically important component of science! Scientists constantly evaluate and criticize evidence. Eventually they come to a consensus about some important questions—such as how heredity makes natural selection possible—so they don't need to keep revisiting those questions. They delve deeper into aspects about the theory of evolution and continue on with the debate.

Great Chain of Being and the March of Progress
Evolution does not produce any kind of hierarchy in life—life does not proceed from simple to complex with humans holding the top spot. Think about it this way. Your family tree was not a chain of being with you as the final outcome:

This Is NOT Your Family Tree

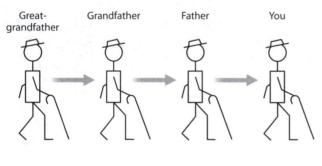

Great-grandfather   Grandfather   Father   You

This Is Your Family Tree

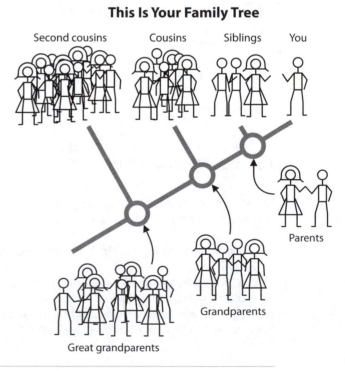

Second cousins   Cousins   Siblings   You

Parents

Grandparents

Great grandparents

So why isn't this evolution?

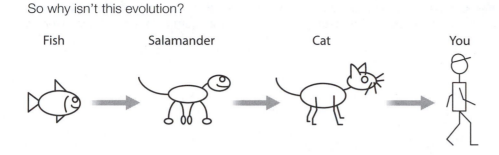

Fish   Salamander   Cat   You

(Created by Matthew Bonnan of Macomb, Illinois, for the Florida Citizens for Science.)

Contemplate

Why did scientists' understanding of nature change as they learned more about the natural world?	Why were Thomas Malthus's ideas about population growth so important to Darwin's and Wallace's ideas about natural selection?

Delve Deeper:

1. What was the most valuable contribution Carl Linnaeus made to scientists today?

2. How did Georges Cuvier combine the geologic discoveries and theories of James Hutton and William Smith and his own observations to decide that geologic formations from very different locations were from the same time period?

3. Why might Lamarck's view of inheritance seem reasonable?

4. Why is Charles Darwin's name more associated with the theory of natural selection than Alfred Russel Wallace's, even though they both developed the idea?

5. What is a major component of the theory of evolution by natural selection that has been added since Darwin's time?

Go the Distance: Examine the Primary Literature

Catherine Powers and David Bottjer at the University of California found that many deep-water bryozoan species went extinct first during the mass die-offs, and then shallow ones disappeared afterward—a pattern that would be predicted if these extinctions were caused, in part, by the disappearance of oxygen from the ocean.

- What are bryozoans?

- What was the evidence?

- Why do the scientists suggest that these patterns suggest long-term changes in the ocean, rather than extraterrestrial mechanisms?

Powers, C. M., and D. J. Bottjer. 2007. Bryozoan Paleoecology Indicates Mid-Phanerozoic Extinctions Were the Product of Long-Term Environmental Stress. *Geology* 35: 995–98. doi:10.1130/G23858A.1.

Test Yourself

1. What are tongue stones?
   a. Fossilized tongues of extinct animals.
   b. Fossilized shark's teeth.
   c. Stones shaped like tongues.
   d. Mythical stones from the 1600s.
   e. None of the above.

2. Who first proposed that life could change over time?
   a. Charles Darwin.
   b. Georges Buffon.
   c. Carl Linnaeus.
   d. Alfred Russel Wallace.
   e. Jean-Baptiste Lamarck.

3. What evidence did James Hutton use to reason that the Earth must be vastly old?
   a. The slow transformation of landscapes caused by erosion.
   b. The rich diversity of fossil species.
   c. William Smith's maps of fossil strata.
   d. Volcanoes that radically transformed the Scottish landscape.
   e. The patterns of fossils of different ages found throughout England.

4. Why did Georges Cuvier reject Lamarck's idea that species were not fixed?
   a. Because he and Lamarck were competitors at the National Museum of Natural History in Paris.
   b. Because Lamarck did not believe in extinction.
   c. Because his experience indicated that huge gulfs existed among major groups of animal fossils with no intermediate fossils.
   d. Because his experience with fossils indicated that all species were created at the same time.
   e. Because he believed the Earth was not old enough for species to change.

5. What would Jean-Baptiste Lamarck and Charles Darwin have agreed upon?
   a. One generation can pass on their traits to the next.
   b. Individual animals and plants can adapt to their environment.
   c. Life was driven from simplicity to complexity.
   d. Both a and b.
   e. All of the above.

6. What is a correct definition of homology?
   a. Common traits due to shared inheritance from a common ancestor.
   b. Common function of traits due to similar usage.
   c. Structure of limbs that are common among all mammals.
   d. All of the above.
   e. None of the above.

7. What did Darwin learn about natural selection from pigeon breeders?
   a. Breeders could select a few birds from each generation with the traits they desired.
   b. This selective breeding would cause the traits to get more and more exaggerated.
   c. Given enough time, this selection could produce new types of body parts, from wings to eyes.
   d. Selection by pigeon breeders is similar to natural selection.
   e. All of the above.
   f. None of the above.

8. Have Darwin's contributions to evolution theory changed since he first proposed the concept of natural selection?
   a. Yes. Scientists now question whether natural selection actually occurs at all.
   b. Yes. Scientists have refuted many of Darwin's ideas, but his fundamental reasoning is still an important contribution.
   c. No. Scientists still believe natural selection operates in the same way that Darwin proposed.
   d. No. Darwin's contributions were never considered that significant.

9. How would you define a scientific theory?
   a. A guess based on a few facts that scientists try to prove as correct.
   b. An educated guess based on some experience that allows scientists to test evidence.
   c. A set of laws that define the natural world.
   d. A set of mechanisms or principles that explain a major aspect of the natural world.
   e. A belief that scientists try to prove as fact.

10. Why do scientists overwhelmingly accept the theory of evolution?
    a. Because of the vast evidentiary support for the theory.
    b. Because the theory explains and predicts independent lines of evidence.
    c. Because scientists have tested and retested the theory's predictions.
    d. Because a scientific theory is more than just an educated guess or vague hunch.
    e. All of the above.

Answers for Chapter 2
Check Your Understanding

1. How can evolution be defined?
   a. Evolution is any change in the inherited traits of a population that occurs from one generation to the next.

      Correct, but so are other answers. Evolution is defined throughout *The Tangled Bank*, but one way to think about the concept is change in the inherited traits *in a population* from one generation to the next. Organisms inherit traits from their ancestors through DNA. Mutations to the DNA underlying those traits can help or hurt the reproductive success and survival of individuals carrying those mutations. More successful individuals tend to leave more surviving offspring with those mutations than less successful individuals, so the number of individuals *within the population* with traits inherited from successful parents will change from one generation to the next (see Chapter 1).

   b. Descent with modification

      Correct, but so are other answers. Evolution is defined throughout *The Tangled Bank*, but Chapter 1 introduces the idea of descent with modification. Organisms inherit traits from their ancestors through DNA. Mutations to the underlying DNA can be harmful or beneficial (modifications), and patterns of reproduction and survival (descent) can affect how common or rare those mutations become in the next generations—hence, descent with modification. Evolutionary biologists study the patterns of modification and the mechanisms of descent.

   c. Mutated genes becoming more or less common over the course of generations.

      Correct, but so are other answers. Evolution is defined throughout *The Tangled Bank*, but one way to think about it is mutations becoming more or less common over the course of generations (see Chapter 1). Organisms inherit traits from their ancestors through DNA. Mutations to the DNA underlying those traits can help or hurt the reproductive success and survival of individuals carrying those mutations. More successful individuals tend to leave more surviving offspring with those mutations than less successful individuals, so the mutated genes may become more or less common in the population over the course of generations.

   d. All of the above are possible definitions.

      Correct. All of these definitions address the genetic changes that occur within lineages over time. Evolution is defined throughout *The Tangled Bank*, but Chapter 1 introduces the underlying concepts.

   e. None of the above is an appropriate definition.

      Incorrect. Each of the definitions above addresses the genetic changes that occur within lineages over time. Evolution is defined throughout *The Tangled Bank*, but Chapter 1 introduces the underlying concepts.

2. What kinds of evidence *haven't* scientists used to study the evolution of whales?
   a. Evidence from DNA.

      Incorrect. In fact, DNA from living mammals provided evidence that whales were related to artiodactyls. Scientists compared snippets of DNA to determine patterns and understand historical relationships. These patterns led to the hypothesis that whales were most closely related to hippos—a hypothesis that could be tested with new fossil evidence (see Chapter 1).

   b. Evidence from fossils, such as their anatomical traits.

      Incorrect. Although Darwin proposed that whales descended from land mammals, very few fossil whales had been discovered to test that hypothesis. Darwin based his hypothesis on evidence from living animals—their similarities (e.g., milk production) and differences (e.g., gills). As more and more fossil whales were discovered, scientists tested Darwin's hypothesis about whale evolution and began asking *how* whales evolved from land mammals (see Chapter 1).

   c. Evidence from geology.

      Incorrect. Geology provides important information about the origin and ages of different rocks. Evolutionary biologists can use this information to understand what kind of habitats the fossils formed in. For example, *Pakicetus* fossils were found in rocks that formed from the sediments of shallow streams that flowed seasonally through hot, dry landscapes—not in rocks formed in ocean habitats. *Pakicetus* probably lived on land, even though it has traits shared by modern whales (see Chapter 1).

d. Evidence from living species.
Incorrect. Evolutionary biologists often use evidence from living species to understand the past. Clues can come from many sources: from how organisms behave to their DNA. Patterns of DNA from cows, goats, camels, and hippos strongly suggested that cetaceans were indeed related to artiodactyls—they shared a common ancestor. And that common ancestor most closely linked whales and hippos. So using living species, scientists could develop a hypothesis about similarities between fossil whales and hippos, and they were able to go out and test that prediction with new evidence from fossil whales (see Chapter 1).

e. All of the above.
Correct. Scientists have used DNA, fossils, geology, and living species to examine the evolution of whale lineages (see Chapter 1). This evidence for evolution is presented throughout *Making Sense of Life*.

3. What trait would you consider the most important in determining whether a fossil should be considered a cetacean or not.

a. The teeth because modern whales have peg-like teeth but *Dorudon*, one of the earliest cetacean fossils completely adapted to life in the water, had diverse teeth.
Incorrect. The teeth of fossil whales can be used to understand some of the relationships among early species, such as between *Dorudon* and *Pakicetus*, but the teeth of *Dorudon* are actually much more similar to those of land mammals than of living cetaceans, even though *Dorudon* was adapted to life in the water. The traits most important to determining membership in groups like the cetaceans should be traits that evolved in the immediate common ancestor of the group and inherited by all its descendants (see Chapter 1).

b. The involucrum because it has a unique form in cetaceans not found in other mammals.
Correct. The traits most important to determining membership in groups like the cetaceans should be traits that evolved in the immediate common ancestor of the group and inherited by all its descendants. Only in cetaceans does the involucrum form a thick lip made of dense bone. *Dorudon*, a genus completely adapted to life in the water, has the same dense, thick form found in modern cetaceans, and so does *Ambulocetus*, a genus with stubby legs that may have been capable of walking (see Chapter 1).

c. The astragalus because if cetaceans evolved from artiodactyls, then early cetaceans should have double-pulley astragali.
Incorrect. The fact that early cetaceans bear a double-pulley astragalus similar to artiodactyls indicates a shared common ancestor, rather than distinguishing cetaceans from artiodactyls. The traits most important to determining membership in groups like the cetaceans should be traits that evolved in the immediate common ancestor of the group and inherited by all its descendants (see Chapter 1).

d. The presence of flippers because the earliest whale fossils had flippers much like those of modern whales.
Incorrect. Although the earliest whale fossils found did have flippers much like those of modern whales, inclusion in the cetaceans does not depend on the presence of flippers. After all, other species wholly unrelated to whales have flippers. The traits most important to determining membership in groups like the cetaceans should be traits that evolved in the immediate common ancestor of the group and inherited by all its descendants (see Chapter 1).

e. The absence of legs because cetaceans do not have legs.
Incorrect. The absence of legs does not necessarily distinguish cetaceans from other mammals. In fact, *Dorudon*, *Pakicetus*, and *Ambulocetus* are all cetaceans that bore legs. The traits most important to determining membership in groups like the cetaceans should be traits that evolved in the immediate common ancestor of the group and inherited by all its descendants (see Chapter 1).

Identify Key Terms

artificial selection	A mechanism that can lead to evolution because humans choose which individuals survive and reproduce because they have specific traits that are preferred for economic or aesthetic reasons.
common descent	The theory that all species on Earth are related, like cousins in a family tree.
extinction	The permanent loss of a population or species, arising with the death or failure to breed of the last individual.
homologies	Traits that are similar because they are inherited from a common ancestor.
natural selection	A mechanism that can lead to evolution because differences in individuals cause some of them to survive and reproduce more effectively than others.
paleontology	The study of prehistoric life.
stratigraphy	The study of layering in rock (stratification) as a method for reconstructing the past.
taxonomy	The science of describing, naming, and classifying species of living or fossil organisms.
theory	An overarching set of mechanisms or principles that explain a major aspect of the natural world.

Homologies:

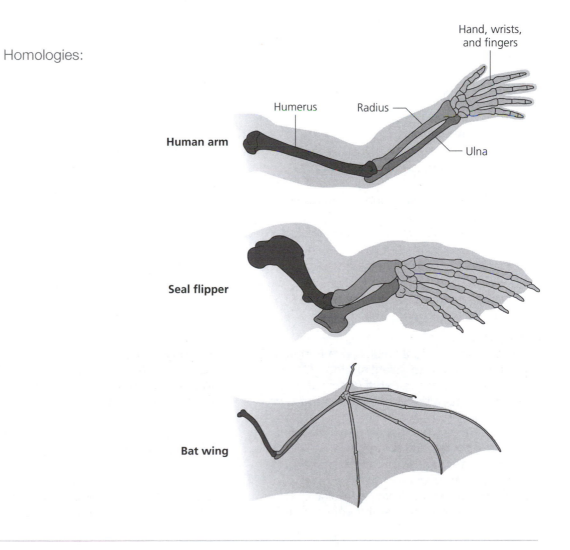

Link Concepts

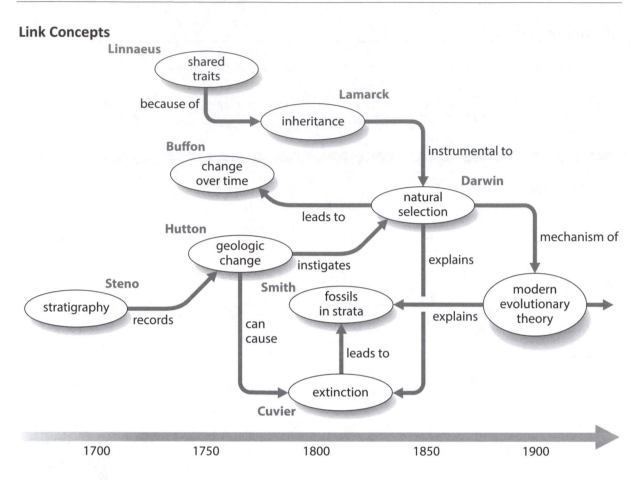

Interpret the Data

- At which stage of human development does the pattern of blood vessels look most similar to the pattern found in sharks?
  29 days.
- If scientists were only to examine human embryos at 56 days, would the homologies be apparent?
  No. By 56 days, the course of human development has altered the arrangement of blood vessels so that the looping pattern apparent early in development is no longer evident.
- What does this homology indicate about the relationship between fishes and humans?
  The fact that fishes and humans share this important homology indicates that we share a common ancestor.

Delve Deeper

1. What was the most valuable contribution Carl Linnaeus made to scientists today?
   Linnaeus's most valuable contribution was the idea that all of life could be organized in a single hierarchy. While he did believe species could sometimes change or bring about new hybrid species, he thought that most species were created in their present form and did not undergo evolution.
2. How did Georges Cuvier combine the geologic discoveries and theories of James

Hutton and William Smith and his own observations to decide that geologic formations from very different locations were from the same time period?
Hutton's view of a world slowly changing over vast periods of time led to Smith's observation that layers of rock could be identified by the kinds of fossils that were found in them. Cuvier studied these fossils to identify the age of the rocks found in different locations and to demonstrate that these

distant rock layers were deposited at the same time.

3. Why might Lamarck's view of inheritance seem reasonable?

Because it's easy to see how we, as humans, can change within our lifetimes. We can grow hair—change our hairstyles to fit the fashion. But we don't pass on our hairstyles to our offspring. We can develop certain muscles required for our lifestyles, like a WWE champion wrestler, but those changes acquired during a wrestler's lifetime aren't passed down to his or her offspring.

4. Why is Charles Darwin's name more associated with the theory of natural selection than Alfred Russel Wallace, even though they both developed the idea?

Darwin first came up with idea after years researching, experimenting, and building support for the theory. Although letters from both Wallace and Darwin were read at the Linnean Society and later published in the society's scientific journal, Darwin's arguments were developed in far more detail. And it was the publication of *On the Origin of Species* in 1859 that brought this mountain of evidence to the world.

5. What is a major component of the theory of evolution by natural selection that has been added since Darwin's time?

Darwin knew nothing about genetic molecules (genes on chromosomes) and their inheritance. The discovery of genetics not only confirmed Darwin's observations, it also added new understanding to the theory of evolution.

Test Yourself

1. b; 2. b; 3. a; 4. c; 5. a; 6. a; 7. e; 8. b; 9. d; 10. e

Chapter 3
WHAT THE ROCKS SAY: HOW GEOLOGY AND PALEONTOLOGY REVEAL THE HISTORY OF LIFE

Check Your Understanding

1. What is a scientific theory?
   a. A belief that scientists try to prove as fact.
   b. A set of laws that define the natural world.
   c. A set of mechanisms or principles that explain a major aspect of the natural world.
   d. A guess based on a few facts that scientists try to prove as correct.
   e. An educated guess based on some experience that allows scientists to test evidence.

2. What was Georges Cuvier's contribution to evolutionary theory?
   a. He strongly objected to Jean-Baptiste Lamarck's ideas about life evolving from simple to complex, drumming Lamarck out of the scientific establishment.
   b. He invented paleontology.
   c. He recognized that some fossils were both similar to and distinct from living species, and many fossil animals no longer existed.
   d. He was the first to organize and map strata according to a geological history.
   e. He rejected the idea that species evolved, instead believing that life's history was a series of appearances and extinctions of species.

3. In the mid-1700s, Carl Linnaeus developed the system for classifying living organisms into nested hierarchies still in use today (e.g., humans belong to the class Mammalia, order Primates, family Hominidae, genus *Homo*, and species *Homo sapiens*). Why are homologies important to that classification system?
   a. Because homology refers to the similarities that different organisms share, such as the arrangement of bones in the limbs of bats, humans, and seals.
   b. Because the nested hierarchies Linnaeus developed were based on patterns of shared traits.
   c. Because homology indicates common ancestry.
   d. All of the above.

Learning Objectives for Chapter 3:

➤ Discuss the importance of radiometric dating to evolutionary theory.
➤ Describe how behaviors observed today can be used to understand plants and animals of the past.
➤ Describe the earliest forms of life on Earth.
➤ Discuss major transitions in the fossil record prior to the appearance of human beings.

Identify Key Terms

Connect the following terms with their definitions on the right:

archaea	Member of a diverse phylum of animals (including humans) that have a notochord, a hollow nerve cord, pharyngeal gill slits, and a post-anal tail as embryos.
bacteria	A lineage of tetrapods that emerged 300 million years ago and gave rise to mammals. This group is distinguished from other tetrapods by the presence of a pair of openings in the skull behind the eyes, known as the temporal fenestrae.
biomarker	A diverse group of animal species that looked like fronds, others like geometrical disks, and still others like blobs covered with tire tracks that existed between 575 and 535 million years ago.
chordate	Humans and all species more closely related to humans than to chimpanzees.
Ediacaran fauna	One of two domains of life that lack a cell nucleus or any other membrane-bound organelles resembling bacteria but distinguished by a number of unique biochemical features.
eukaryotes	A tool based on changes in the ratios of isotopes over time that allows scientists to estimate the ages of fossils.
fossilization	Layered structures formed by the mineralization of bacteria.
hominin	Molecular evidence of life in the fossil record, such as fragments of DNA, molecules such as lipids, or isotopic ratios.
isotope	A domain of life characterized by unique traits including membrane-enclosed cell nuclei and mitochondria, including animals, plants, fungi, and protists.
radiometric dating	The process of preserving remains of dead animals and other organisms through mineralization over extremely long periods of time.
stromatolite	One of two domains of life that lack a cell nucleus or any other membrane-bound organelles, including organisms such as *E. coli* and other familiar microbes.
synapsid	Microscopic crystals that preserve the oldest known carbon isotopes that can provide evidence of the Earth's very early history.
zircon	An alternative form of a particular element that shares the same number of protons but differs in the number of neutrons.

The Tangled Bank Study Guide

Link Concepts

1. Deep time is pretty abstract; it's hard to imagine just how long 1 million years ago is, let alone 4.567 billion years! If you could count one number every second, it would take just over 11.5 days to count to a million (that's without eating, sleeping, or taking any breaks). It would take almost 32 *years* (31 years and 255 days to be exact) to count to a billion! At that rate, you couldn't even count to 4.567 billion years in a lifetime.

   It's easier to get an idea of the *relative* age of events in the Earth's past. Try this demonstration using a 1000-sheet roll of toilet paper. (You can use a roll with fewer sheets—just divide the number in the "Years before Present" column by the total number of sheets in the roll.) The roll represents the entire age of the Earth, and you can mark important events (such as the oldest known zircons, or the first multicellular organisms) on different sheets as they are rolled out. The more room you have to spread out, the more you'll get a feel for exactly how short a time humans have been around on this planet!

Event	Years before Present	Number of Sheets
Origin of earth	4,567,000,000	1,000.000
Oldest known zircons	4,404,000,000	964.309
Rocks containing carbon	3,700,000,000	810.160
Archaea (biomarker)	3,500,000,000	766.367
Stromatolites	3,450,000,000	755.419
Cyanobacteria	2,600,000,000	569.302
Multicellularity	2,100,000,000	459.820
Eukarya	1,800,000,000	394.132
Algae	1,600,000,000	350.339
Red algae	1,200,000,000	262.755
Green algae	750,000,000	164.222
Sponges	650,000,000	142.325
Worm-like fossil traces	585,000,000	128.093
Ediacaran Fauna	575,000,000	125.903
Cambrian period	542,000,000	118.677
Chordates	515,000,000	112.765
Oldest invertebrate traces	480,000,000	105.102
Oldest plant fossils	475,000,000	104.007
Oldest insect fossils	400,000,000	87.585
Wattieza fossil	385,000,000	84.300
Millipede fossil	428,000,000	93.716
Oldest vertebrate traces	390,000,000	85.395
Oldest tetrapods	370,000,000	81.016
Synapsids	320,000,000	70.068
Permian extinction	252,000,000	55.178
First dinosaurs	230,000,000	50.361
Mammals & birds	150,000,000	32.844

Oldest flowering plants	132,000,000	28.903
Grasses	70,000,000	15.327
Cretaceous extinction	66,000,000	14.451
Primates	50,000,000	10.948
Ambulocetus	49,000,000	10.729
Dorudon	40,000,000	8.758
Earliest ape fossils	20,000,000	4.379
Sahelanthropus	7,000,000	1.533
Human-chimpanzee split	6,600,000	1.445
Australopithecus afarensis	3,850,000	0.843
Oldest stone tools	2,600,000	0.569
Homo erectus	1,890,000	0.414
Homo heidelbergensis	600,000	0.131
Homo neanderthalensis	400,000	0.088
Humans	200,000	0.044
Present	-	-

2. Develop a concept map that shows the relationships among different types of evidence scientists use to understand the ancient past. Some concepts you might include are radiometric dating, stratigraphy (the study of layering in rock), DNA, modern studies of behavior, comparative anatomy, microscopic structures.

Interpret the Data

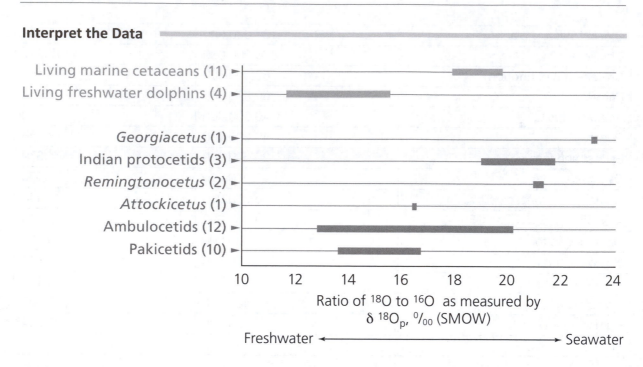

Figure 3.7 shows the ranges of oxygen isotopes extracted from the teeth of fossil whales that lived during the Eocene. Ratios of the two most common isotopes of oxygen in nature, ^{18}O and ^{16}O, are listed along the x axis. Lower ratios (toward the left side of the x axis) indicate a freshwater environment, and higher ratios (toward the right the side of the x axis) indicate a marine environment. Fossils are listed from oldest (Pakicetids) up to youngest (*Georgiacetus*) along the y axis.

Which fossils likely lived in or near freshwater?

Which fossils likely lived in or near seawater?

Which fossils are older—those that lived in or near freshwater or those that lived in or near seawater?

How would you interpret the evolution of whales based on this information?

Games and Exercises

A Sweet "Half Life"
Understanding decay rates can be difficult because the exact time that a certain nucleus will decay can't be predicted. Decay rates can be measured, however, and you can get a feel for how those measurements are determined by examining the "decay rates" of M&Ms or Skittles (or any candy with two "sides").

Use the blank side of the candy to represent the radioactive nuclei that have not lost a neutron and the marked side to represent nuclei that have lost a neutron and become a different isotope of the candy "element." On average, every 100 years one-half of the "blank" candies present will decay (lose a neutron) to become the "marked" isotope, so the half-life of the candy element is 100 years. Count out 80 candies and place them in a cup. To simulate decay, simply empty the candies out onto a flat surface. If the candy lands with the blank side up, it has not decayed, but if lands with the marked side up, it has decayed. Count all the candies that have decayed and enter your observations in a table starting with the first simulation, i.e., the first 100 years. Record the number of blank candies and the marked candies in a data table. Try to predict the number of candies that will decay in the next 100 years (i.e., the next time you empty the cup on the table).

Simulation	Blank candies	Marked candies	Ratio of blank-to-marked candies	Prediction for next simulation
1 (100 years)				
2 (200 years)				
3 (300 years)				
4 (400 years)				

Remove all the "marked" isotopes (you can eat them if you're hungry), and repeat the simulation using only the remaining radioactive "nuclei" (the blank candies). Continue this process until there are no radioactive isotopes left. Add more rows to the data table as needed.

Plot the data on a graph with the number of years that passed on the x axis and the number of radioactive nuclei on the y axis.
- Could you predict how many candies decayed after each simulation?
- What do you think would happen to your graph if you had more data?
- About how many nuclei would decay if you started with twice as many candies? What about if you started with half as many candies? What does that tell you about how the quantity of "radioactive isotopes" affects the number that decay?
- How is radioactive decay sort of like gambling or playing the lottery?
- If scientists can't know *exactly* how many nuclei decayed, why do they use radiometric dating as a tool?

This activity is based on Radioactive M&Ms by Darren Fix at ScienceFix.com (http://www.sciencefix.com/).

You can watch this simulation unfold in a video from ScienceFix.com. The video comes with a downloadable instruction sheet for understanding radiometric dating.

http://www.sciencefix.com/home/2010/4/8/lesson-m-m-radiometric-dating.html

Go Online

 How the Earth Was Made

From the *History Channel*, this short clip discusses Lord Kelvin's ideas about the Earth and the discovery of radiometric dating.

http://www.history.com/shows/how-the-earth-was-made/videos/the-age-of-earth#the-age-of-earth

 What are Isotopes?

Tyler DeWitt does a great job explaining exactly what isotopes are by comparing them with the different models of automobiles.

http://www.youtube.com/watch?v=EboWeWmh5Pg

 Virtual Dating

This website goes into a bit more detail about radiometric dating, compete with quizzes to test yourself.

http://www.sciencecourseware.org/virtualdating/

 RareResouce.com

Shawn Mike, a graphic designer from Chicago, developed RareResource. The website includes a wealth of information and some videos on dinosaurs.

http://www.rareresource.com/dinosaur_videos.htm

Did Dinosaurs Travel in Herds or Packs?

The American Museum of Natural History discusses how trackways can provide important clues to the behavior of birds and their dinosaur relatives.

http://www.amnh.org/exhibitions/past-exhibitions/dinosaurs-ancient-fossils-new-discoveries/trackways

 Watch a quick video about the evidence for herds here:
http://www.youtube.com/watch?v=jtbpusl0Vo0

Paleontologists Teach Medical Students about Fossil Tumors

The oldest evidence of cancer in the fossil record may provide valuable insight to the evolution of this disease in humans. So modern-day behaviors can help paleontologists understand the past, and paleontologists can help modern-day medical doctors understand the future. Check out this site by *Science Daily*. It includes a short video about how important understanding evolution is to modern medicine.

http://www.sciencedaily.com/videos/2006/0607-jurassic_docs.htm

---

 100 Greatest Discoveries: Laetoli Footprints

This video from *HowStuffWorks* shows one of the 100 greatest discoveries—the discovery of fossilized human footprints.

http://videos.howstuffworks.com/science-channel/29288-100-greatest-discoveries-laetoli-footprints-video.htm

---

 Dating the Chicxulub Impact

Did dinosaurs go extinct as a result of a large asteroid impact? Scientists are honing in on an answer. Paul Renne and his colleagues present new evidence using highly precise radiometric dates for the stratigraphic layers surrounding the Cretaceous-Paleogene boundary that indicates the impact helped along an already stressed global ecosystem.

http://www.sciencemag.org/content/327/5970/1214

For a more recent discussion of the research, check out this Perspective by Heiko Pälike.

http://www.sciencemag.org/content/339/6120/655

Common Misconceptions

Age of the Earth

Scientists no longer debate the overall age of the Earth—the 4.567-billion-year age is widely accepted simply because of the quantity of evidence from many lines of evidence that supports that age. Scientists certainly continue to debate specific aspects of the formation of the Earth, such as how mineral grains condensed and aggregated to form the Earth.

Radiometric Dating

A lot of misinformation about radiometric dating exists, but the process is a valid scientific tool that has been tested and retested by scientists worldwide. Just because radiometric dating provides a range of ages doesn't mean it's inaccurate. Rocks contain many different kinds of isotopes, and scientists consistently derive the same ages of rocks using decay rates of different isotopes. Radiometric dates are not only supported by volumes of consistent results, they have also been validated with independent lines of evidence, such as tree rings and varved sediments (for more on varved sediments, see Hughen and Zolitschka 2007, http://www.sciencedirect.com/science/article/pii/B0444527478000648).

Contemplate

Why is understanding how organisms fossilize important to paleontology?	How certain are evolutionary biologists about the age of the earliest life on Earth?

Delve Deeper

1. What evidence refutes Kelvin's claim that the Earth is only 20 million years old?

2. Why do scientists consider radiometric dating a valid way to measure the age of rocks?

3. What are stromatolites, and why are they evidence of very early life?

4. Why are fossils rare?

5. How have fossils allowed scientists to better understand the behavior of extinct animals?

6. Why do evolutionary biologists believe that plants and fungi may have been integral to each other's colonizing dry land?

7. What are tetrapods? What is the oldest evidence we have of tetrapods?

Go the Distance: Examine the Primary Literature

Fossilized dung? Yes, dung can fossilize, too, under the right conditions. Now, fossilized dinosaur dung, known as a coprolite, is giving scientists insight into dinosaur diets. Dolores Piperno and Hans-Dieter Sues discuss recent evidence that suggests not only that dinosaurs ate grasses, but also that grasses originated and diversified right along with dinosaurs. These two groups may be an ancient example of the evolutionary interactions between plants and the animals that eat them.

- Does this article explain the results of the authors' original research?

- What is the purpose of this article?

- Do the authors provide additional evidence to support their claims?

Piperno, D. R., and H.-D. Sues. 2005. Dinosaurs Dined on Grass. *Science* 310 (5751): 1126–28. doi:10.1126/science.1121020. http://www.sciencemag.org/content/310/5751/1126.summary.

Test Yourself

1. Why was Lord Kelvin's estimate for the age of the Earth so inaccurate?
   a. Because he did not account for the dynamic nature of the Earth's crust.
   b. Because he could not directly measure the age of any rock.
   c. Because he used a model to generate a prediction for the age of the Earth.
   d. Because he was trying to challenge evolution by natural selection.
   e. Because he could not estimate cooling of rocks found deeper than in mines.

2. Which of the following molecules has not been used as a biomarker, an organic chemical signature of once living organisms?
   a. Carbon isotopes.
   b. Oxygen isotopes.
   c. Sodium ions.
   d. Okenane.
   e. All of the above have been used as organic chemical signatures.

3. What are stromatolites?
   a. Layered structures formed by the mineralization of bacteria.
   b. Bacterial biomarkers.
   c. Fossil bacteria.
   d. All of the above.
   e. None of the above.
4. What evidence supports the idea that fungi were important to the establishment of land plants?
   a. Modern land plants live in close association with fungi.
   b. Scientists have found fossils of fungi intermingled with fossils of plants.
   c. The oldest fossils of fungi are groups that provide nutrients to plants.

   d. All of the above provide evidence to support this claim.
   e. None of the above provides evidence to support this claim.

5. How would you describe the fossil record?
   a. A complete record of the history of life.
   b. An accurate record of recent fossils only.
   c. A patchy, fragmented record that does not include transitional species.
   d. A patchy, fragmented record that indicates that humans are relatively recent arrivals.
   e. Very patchy, fragmented, and without foundation for evolutionary theory.

6. Which of the following statements about the history of life is <u>true</u>?
   a. Because cyanobacteria, the lineage of bacteria that carries out photosynthesis, have not changed in 1.6-billion years, evolution must not be occurring.
   b. Scientists have not found any fossils from before the Cambrian period, 542 million years ago.
   c. Scientists have not been able to resolve many of the relationships among the three great branchings, so they cannot know anything about the early history of life.
   d. Living bacteria can offer clues about how the first multicellular animals evolved.
   e. None of the above are true statements.

Answers for Chapter 3
Check Your Understanding

1. What is a scientific theory?
   a. A belief that scientists try to prove as fact.
      Incorrect. Scientists try to avoid letting their beliefs interfere with their work. More importantly, science is not about proving anything as fact. Science is about testing theories, weight of evidence, and ultimately understanding and explanation. Facts are simply consistent observations. Evolution is a testable theory that provides an overarching set of mechanisms or principles that explain a major aspect of the natural world.
   b. A set of laws that define the natural world.
      Incorrect. Laws can be important in science—they define relationships. Laws describe consistent observations, but they don't explain those observations. Theories explain relationships. In fact, laws can be essential components of scientific theories. Theories, such as evolution, are an overarching set of mechanisms or principles that explain major aspects of the natural world.
   c. A set of mechanisms or principles that explain a major aspect of the natural world.
      Correct. A scientific theory is different from the word *theory* used in everyday language. Scientific theories are not just guesses; they are overarching sets of mechanisms or principles that explain and provide testable predictions.
   d. A guess based on a few facts that scientists try to prove as correct.
      Incorrect. Although in everyday language, a theory is often considered a guess, in science a theory is not a guess. A theory, such as evolution, is an overarching set of mechanisms or principles that explains a major aspect of the natural world. More importantly, science is not about facts or proving ideas as correct. Science is about testing theories, weight of evidence, and ultimately understanding and explanation.
   e. An educated guess based on some experience that allows scientists to test evidence.
      Incorrect. The idea of an educated guess is closer to the concept of a hypothesis—a tentative explanation grounded in some evidence. Many theories can develop from hypotheses as the overarching mechanisms and principles that guide the explanation and

offer predictions are developed and tested. Evolutionary theory initially developed from early hypotheses about fossils and their geologic relationships; Darwin contributed the first overarching set of mechanisms as natural selection and sexual selection.

2. What was Georges Cuvier's contribution to evolutionary theory?
   a. He strongly objected to Jean-Baptiste Lamarck's ideas about life evolving from simple to complex, drumming Lamarck out of the scientific establishment.
      Incorrect. Cuvier certainly disagreed with Lamarck's theory, but his contribution to evolutionary theory is not that he successfully destroyed the career of a scientist with whom he disagreed. Scientists regularly disagree, often until the balance of evidence weighs in favor of one theory or another. Cuvier's contributions to evolutionary theory come from his work on comparative anatomy and the concept of extinction.
   b. He invented paleontology.
      Incorrect. The science of paleontology was not invented but developed over years as scientists began to uncover and study fossils. Cuvier contributed significantly to the science of paleontology through his exhaustive anatomical studies of fossils; however, many individuals were contributing to the science of paleontology prior to Cuvier's work.
   c. He recognized that some fossils were both similar to and distinct from living species, and many fossil animals no longer existed.
      Correct. Cuvier studied the fossil remains of elephants and compared them to living elephants. He discovered that some characters of the fossils, such as the shapes of teeth, were distinct from living elephants and provided some of the first compelling evidence for extinction.
   d. He was the first to organize and map strata according to a geological history.
      Incorrect. William Smith discovered that layers of rocks contained distinctive groups of fossils, and he was able to organize strata into a geological history as a result. Cuvier used the system developed by Smith to map strata in other parts of the world.
   e. He rejected the idea that species evolved, instead believing that life's history was a

series of appearances and extinctions of species.

Incorrect. Cuvier believed that life's history was a series of appearances and extinctions of species rather than accepting the concept of evolution. His contributions were significant to the developing field of paleontology, however.

3. In the mid-1700s, Carl Linnaeus developed the system for classifying living organisms into nested hierarchies still in use today (e.g., humans belong to the class Mammalia, order Primates, family Hominidae, genus *Homo*, and species *Homo sapiens*). Why are homologies important to that classification system?

a. Because homology refers to the similarities that different organisms share, such as the arrangement of bones in the limbs of bats, humans, and seals.

Correct, but so are other answers. Homology does refer to shared similarities among organisms, but homology is also important to Linnaean taxonomy because of its use in the development of nested hierarchies and its indication of common ancestry.

b. Because the nested hierarchies Linnaeus developed were based on patterns of shared traits.

Correct, but so are other answers. Homology is important to the development of nested hierarchies because it refers to the shared traits among organisms, such as the arrangement of bones in the limbs of bats, humans, and seals. Homology also is evidence of common ancestry.

c. Because homology indicates common ancestry.

Correct, but so are other answers. Homology is evidence of common ancestry, but its importance to Linnaean taxonomy stems from the patterns of shared traits among different organisms that arise from common ancestry and how those patterns can be grouped into nested hierarchies.

d. All of the above.

Correct. Homology refers to the shared similarities among organisms. Homology is extremely important to Linnaean taxonomy because patterns of homology are the basis for the development of nested hierarchies. More importantly, homology is evidence of common ancestry.

Identify Key Terms

archaea	One of two domains of life that lack a cell nucleus or any other membrane-bound organelles resembling bacteria but distinguished by a number of unique biochemical features.
bacteria	One of two domains of life that lack a cell nucleus or any other membrane-bound organelles, including organisms such as E. coli and other familiar microbes.
biomarker	Molecular evidence of life in the fossil record, such as fragments of DNA, molecules such as lipids, or isotopic ratios.
chordate	Member of a diverse phylum of animals (including humans) that have a notochord, a hollow nerve cord, pharyngeal gill slits, and a post-anal tail as embryos.
Ediacaran fauna	A diverse group of animal species that looked like fronds, others like geometrical disks, and still others like blobs covered with tire tracks that existed between 575 and 535 million years ago.
eukaryotes	A domain of life characterized by unique traits including membrane-enclosed cell nuclei and mitochondria, including animals, plants, fungi, and protists.
fossilization	The process of preserving remains of dead animals and other organisms through mineralization over extremely long periods of time.
hominin	Humans and all species more closely related to humans than to chimpanzees.
isotope	An alternative form of a particular element that shares the same number of protons but differs in the number of neutrons.

radiometric dating	A tool based on changes in the ratios of isotopes over time that allows scientists to estimate the ages of fossils.
stromatolite	Layered structures formed by the mineralization of bacteria.
synapsid	A lineage of tetrapods that emerged 300 million years ago and gave rise to mammals. This group is distinguished from other tetrapods by the presence of a pair of openings in the skull behind the eyes, known as the temporal fenestrae.
zircon	Microscopic crystals that preserve the oldest known carbon isotopes that can provide evidence of the Earth's very early history.

Interpret the Data

- Which fossils likely lived in or near freshwater?
  Pakicetids, *Ambulocetus*, and *Attockicetus*
- Which fossils likely lived in or near seawater?
  Remingtonocetus, Indian protocetids, and *Georgiacetus*
- Which fossils are older—those that lived in or near freshwater or those that lived in or near seawater?
  Fossils that lived in or near fresh water.
- How would you interpret the evolution of whales based on this information?
  The whale lineage was initially associated with fresh water. Over time, the lineage of whales gradually moved into marine environments and became more and more aquatic.

Delve Deeper

1. What evidence refutes Kelvin's claim that the Earth is only 20 million years old?
   Radioactivity was discovered in 1896 (Kelvin died in 1907), and around 1910, scientists had figured out a way to measure the absolute age of rocks using decay rates of isotopes. Using radiometric dating, scientists have shown that the Earth formed as part of a dust cloud around the Sun 4.567 billion years ago. Also, Kelvin didn't know that the Earth's interior was dynamic, with hot rock rising, cooling, and sinking. He assumed the planet was a rigid sphere, so his model didn't account for the greater heat flow that results from this movement.

2. Why do scientists consider radiometric dating a valid way to measure the age of rocks?
   Scientists often spend a lot of time critically examining other scientists' evidence, and radiometric dating has been used as a tool to estimate age for decades. They've tested and retested radiometric dating and compared ages derived from radiometric dating with other methods for dating and found very consistent results.

3. What are stromatolites, and why are they evidence of very early life?
   Stromatolites are ancient rocks found in some of the oldest geological formations on Earth. They bear striking microscopic similarities to large mounds built by colonies of bacteria alive today. Although these formations are rare on Earth today, they are abundant in the early fossil record. So stromatolites are not only evidence of early bacterial life, they also may be important to understanding life on Earth 3.45 billion years ago.

4. Why are fossils rare?
   Fossils are rare for several reasons. Most importantly, not all animals' remains are left to be fossilized. Most animals are food for other organisms, so their bodies may or may not even be available to be fossilized. And then, organisms that die are left to the elements, like wind and rain, which can destroy any remaining evidence of their existence. If an organism is buried by sediments and its remains fossilize, paleontologists have to be able to access the rocks—and access them quickly. Over time the fossils themselves can again be exposed to the elements, destroying what is now pretty rare evidence. Soft tissues, like skin and organs, can be especially difficult because it takes an especially rare set of circumstances to mineralize and preserve these tissues.

5. How have fossils allowed scientists to better understand the behavior of extinct animals?
Scientists have found fossils that exhibit different behaviors, such as live birth, predation, herding, and even parental care. Scientists can examine the behavior of animals today and combine that knowledge with the physical evidence from fossils to develop hypotheses about how the extinct creatures lived. They can test their predictions with other lines of evidence, such as the organization of the melanosomes of fossil feathers. As a result, they can begin to understand how dinosaurs may have used feathers in courtship display, a behavior common in modern birds.

6. Why do evolutionary biologists believe that plants and fungi may have been integral to each other's colonizing dry land?

Because the oldest fossil fungi are found mingled with the early land plant fossils. Today, similar fungi live in very close association with land plants and supply nutrients to the roots of plants in exchange for the organic carbon that the plants create in photosynthesis.

7. What are tetrapods? What is the oldest evidence we have of tetrapods?
Tetrapods are vertebrates with four legs or limbs (literally tetrapod means four footed), or any vertebrate descended from an ancestor with four limbs (so whales and snakes are considered tetrapods). The earliest evidence of tetrapods actually comes from fossilized footprints that were made on a coastal mudflat that later solidified into rock about 390 million years ago. The oldest known fossils are about 370 million years old.

Test Yourself
1. a; 2. c; 3. a; 4. d; 5. d; 6. d

Chapter 4
THE TREE OF LIFE: HOW BIOLOGISTS USE PHYLOGENY TO RECONSTRUCT THE DEEP PAST

Check Your Understanding

1. How does homology relate to the theory of evolution?
   a. Homology refers to traits that superficially look alike, like bat wings and human arms; the theory of evolution cannot explain the differences.
   b. Homology refers to traits that look alike but have entirely different origins, like the fins of dolphins and fish; the theory of evolution cannot explain the origins.
   c. Homology refers to traits that are structurally similar in different organisms, like bat wings and human arms, because they each were inherited from a shared common ancestor with those traits; the theory of evolution provides a mechanism for those observations.
   d. Homology refers to traits that have converged on a shared form; the theory of evolution provides a mechanism for those observations.

2. How do scientists know if their explanations are correct?
   a. They don't always know; they can only rely on the quality and quantity of evidence available to them.
   b. They do an experiment, and if it turns out the way they predicted it would, they know they are correct.
   c. They publish the results of their experiments to prove they are correct.
   d. Unless scientists can directly observe the phenomenon they are interested in, they cannot know whether their explanations are correct.

3. How accurate is radiometric dating?
   a. Not very accurate because scientists cannot know how much of a particular isotope was originally present in a rock, and thus they cannot know how much of it has decayed.
   b. Very accurate because scientists can determine the exact age of rocks and the fossils found within them.
   c. Accurate enough because radiometric dating can determine estimates for the ages of rocks and fossils, often with relatively small margins of error.
   d. Not very accurate because scientists use probabilities to determine decay rates for each isotope they use in radiometric dating.

Learning Objectives for Chapter 4:

➤ Identify the components of a phylogenetic tree.
➤ Explain how phylogenies serve as scientific hypotheses.
➤ Discuss the types of evidence used in developing a phylogeny of tetrapods.
➤ Discuss how a phylogeny can be used to explain a complex adaptation, such as the bones of the middle ear in mammals and their ancestors.
➤ Describe the evidence supporting the hypothesis that feathers evolved before flight.
➤ Analyze the evidence for the evolution of bipedality in the common ancestors of humans.

Identify Key Terms

Connect the following terms with their definitions on the right:

branch	A single "branch" in the tree of life representing an organism and all of its descendants.
clade	A lineage evolving through time between successive speciation events.
common ancestor	A member of the superclass Tetrapoda that includes the first four-limbed vertebrates and their descendants, including amphibians, birds, and mammals.
homology	The similarity of characteristics in different species because of a common ancestor.
hypothesis	The terminal end of an evolutionary tree, representing the species, the molecule, or the population that is to be compared to other species, molecules, or populations.
node	A member of the subclass Theria within the class Mammalia that includes marsupial and placental mammals.
phylogeny	A point in a phylogeny where a lineage splits (a speciation event).
tetrapod	A visual representation of the evolutionary history of populations, genes, or species.
therian	A proposed explanation for an observable phenomenon, usually forming a basis for possible experiments to confirm its viability.
tip	The ancestral form or species from which two or more different forms or species descended.

Link Concepts

1. Fill in the following concept map with the key terms from the chapter:

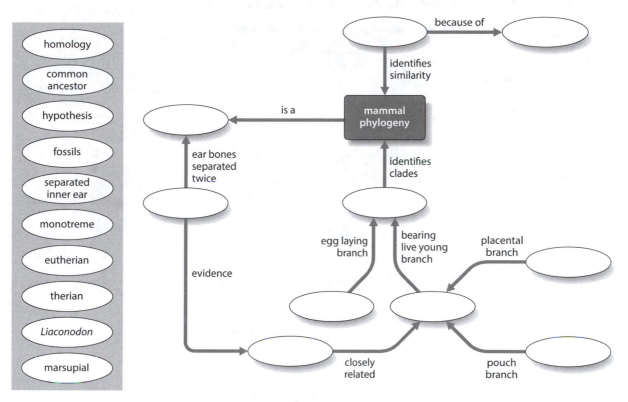

homology

common ancestor

hypothesis

fossils

separated inner ear

monotreme

eutherian

therian

Liaconodon

marsupial

2. Develop your own concept map that explains the relationship between a phylogeny and human evolution.

Interpret the Data

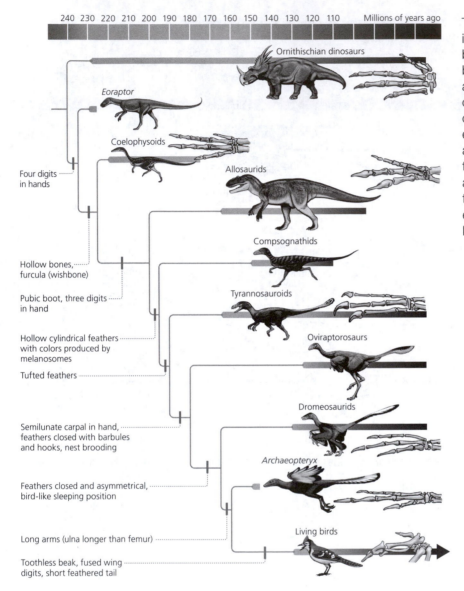

240 230 220 210 200 190 180 170 160 150 140 130 120 110   Millions of years ago

Ornithischian dinosaurs

Eoraptor

Coelophysoids

Four digits in hands

Allosaurids

Hollow bones, furcula (wishbone)

Compsognathids

Pubic boot, three digits in hand

Tyrannosauroids

Hollow cylindrical feathers with colors produced by melanosomes

Oviraptorosaurs

Tufted feathers

Dromeosaurids

Semilunate carpal in hand, feathers closed with barbules and hooks, nest brooding

Archaeopteryx

Feathers closed and asymmetrical, bird-like sleeping position

Living birds

Long arms (ulna longer than femur)

Toothless beak, fused wing digits, short feathered tail

This evogram (Figure 4.15) illustrates the lineage of birds by arranging the branches of a cladogram along a timeline (top). Some of the key character changes that distinguish each branch of theropods are listed at each node. The thick blue lines show the ages of known fossils, and the ages of the nodes are estimates based on the known ages of the fossils.

Explain why living birds are considered dinosaurs.

What trait(s) distinguish living birds from theropod dinosaurs?

Which branches of theropods have feathers?

Which branches may have been able to fly?

Games and Exercises

Phylogenies can be incredibly powerful tools for studying relationships in more than just biological organisms. Practically, the analysis is the same whether you want to look at the relationships among beetles or the evolution of languages; they are simply nested hierarchies—groups within groups within groups.

A cladogram groups organisms based on their shared derived characteristics. A simple example adapted from an ENSI/SENSI lesson plan called "Making Cladograms" (http://www.indiana.edu/~ensiweb/lessons/mclad.html) illustrates how this nesting works.

Characters	Shark	Bullfrog	Kangaroo	Human
Vertebrae	yes	yes	yes	yes
Two pairs of limbs		yes	yes	yes
Mammary glands			yes	yes
Placenta				yes

One way to appreciate the relationships among the organisms in the table above is to make a Venn diagram. Venn diagrams are like nesting doll toys—with the largest on the outside and the smallest on the inside. Start with the character that is shared by all the taxa, and then each successive group is smaller because they share a more limited set of characters.

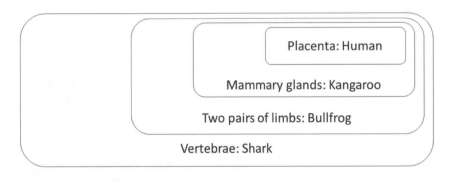

The resulting cladogram looks like this:

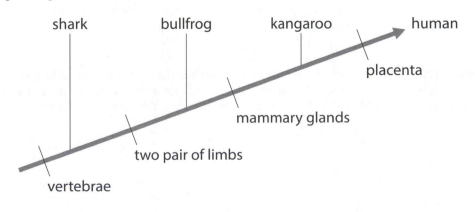

Other analyses can be really complex, but learning how to develop a phylogenetic tree just takes a little practice. Try this exercise with some evidence from stick insects using the shared differences to develop a tree.

Stick insects are an order of insects, many of which have striking adaptations in camouflage. Some species look like sticks; others look like leaves—they even "behave" like leaves, swaying as if blown by a breeze. Within the Phasmatodea, however, classification of more specific taxonomic levels is not well resolved. The Euphasmatodea is essentially a synonym for the order, and taxonomists may divide the order into two or three suborders. In fact, scientists are actively generating new hypotheses about the relationships among this interesting group, especially as new fossils are discovered that can help resolve the history of the major clades.

Here is a simplified cladogram of the extant (living species) stick insects:

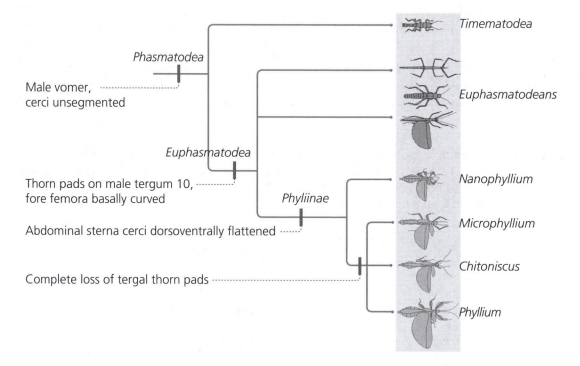

Some helpful definitions:
vomer = A hardened section of the male's 10th abdominal sternum that is used during copulation;
cerci = paired appendages extending from the anal segment of many arthropods;
tergum (plural: terga) = the dorsal part of a segment (see sternum);
femora (single: femur) = the thigh, typically the longest segment of the insect leg; and
sternum (plural: sterna) = the ventral part of a segment (see tergum).

You can use this cladogram to read "backward" and determine the characters used to distinguish each group of stick insects. Fill in the table with the character states for each taxon shown. Write "yes" (or the number 1) for each taxon that has the character state and "no" (or the number 0) for those that don't.

	Vomer present Cerci unsegmented	Thorn pads on male T10 Fore femora curved	Abdominal sterna Cerci dorsoventrally flat	Loss of tergal thorn pads
Timematodea (T)				
Euphasmatodeans (E)				
Nanophyllium (N)				
Microphyllium (M)				
Chitoniscus (C)				
Phyllium (P)				

- Are the Euphasmatodea a monophyletic clade?

Draw a Venn diagram that indicates the nested relationship among the stick insects.

Archipseudophasmatidae   *Eophyllium*

Now consider some fossil evidence. *Timematodea* and *Archipseudophasmatidae* were both found in amber that dates from approximately 40 million years ago. Male *Archipseudophasmatidae* have thorn pads on their 10th terga, and the fore femora are curved—they do not have dorsoventrally flattened cerci or abdominal sterna. *Eophyllium* has abdominal sterna, but it lacks dorsoventrally flattened cerci (Wedman et al. 2007). *Eophyllium* was found in deposits that date to 47 million years ago. Where would you place these two new genera given your current understanding?

Draw a new Venn diagram:

Now draw a new cladogram:

- Did the addition of *Eophyllium* clarify the resolution of the Euphasmatodea?

Tree Challenge:
Develop a phylogenetic tree that illustrates the relationships among your **shoes**

Phylogenies can also be used (cautiously) to ask questions about human artifacts, like the evolution of fashion. To keep it simple, grab one shoe from five different pairs. Identify five different characters that can be used to identify the shoes, such as presence/absence of shoe laces, sole material (natural/man-made), type of shoe (boot, sandal, tennis shoe), toe shape (round/pointed)—you will likely have to come up with your own categories depending on the shoes you choose. Develop a data matrix by listing the characters in columns and the species in rows. Then identify the differences between species pairs in another data matrix. Build the tree by plotting out the differences. It may be difficult, but that's what evolutionary biologists have to deal with every day. They make educated guesses about the relationships—the phylogenetic tree is a hypothesis that they can test with additional data.

As far as shoes go, people are notorious for borrowing other people's ideas (and ideas in general). So the observed "homology" may not really result from "descent"—it might result because one lineage (i.e., one fashion designer) took an idea from another lineage—something akin to horizontal gene transfer (see Box 5.2). Do you think "borrowing" might influence the relationships in the tree?

Go Online

Why Study the Tree of Life?

Why Study the Tree of Life is a 4-minute video produced by the Yale Peabody Museum of Natural History that highlights the role phylogenies play in understanding and responding to challenges we face on this planet, from increasing yields of our food crops to protecting endangered species to curing disease. The site also has a variety of information and other links.

http://archive.peabody.yale.edu/exhibits/treeoflife/film_study.html

What did T. rex *Taste Like?*

by Jennifer Johnson Collins, Judy Scotchmoor, and Caroline Stromberg

The University of California Museum of Paleontology hosts this extensive look into how cladograms can be used to ask interesting and bizarre questions. Going through the exploration takes a bit of time, but wouldn't it be cool to tell your friends you know what Tyrannosaurus rex tasted like?

http://www.ucmp.berkeley.edu/education/explorations/tours/Trex/index.html

The Tangled Bank Study Guide

 Cladograms

In this video, Paul Andersen shows how to construct a simple cladogram from a group of organisms using shared characteristics.

https://www.youtube.com/watch?v=ouZ9zEkxGWg

ScienceShot: Stubby-Tailed Dinosaurs Shook Their Thing

New research about the function of feathers in early theropod dinosaurs.

http://news.sciencemag.org/sciencenow/2013/01/scienceshot-stubby-tailed-dinosa.html?ref=em

---

Tree Thinking Challenge

 Also from *Science*, David A. Baum, Stacey DeWitt Smith, and Samuel S. Donovan developed two quizzes designed to help you understand the phylogenetic relationships.

http://www.sciencemag.org/content/suppl/2005/11/07/310.5750.979.DC1/Baum.SOM.pdf

Common Misconceptions

Reading phylogenetic trees can be difficult because they are not necessarily intuitive. Misunderstanding is easy. Following are several common misconceptions related to reading trees.

The Flow of Time

Trees can be depicted a number of different ways. In all of these figures, A–D represent the species or groups of interest, and the nodes and branches show their historical relationships.

In this style, time is represented from bottom to

top—the extant (living) species or groups are listed at the top, and each successive node down represents an earlier and earlier common ancestor.

In this style, time is just the opposite—the extant

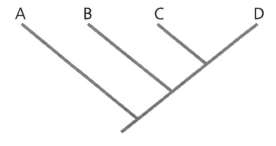

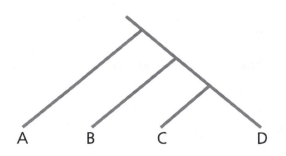

(living) species or groups are listed at the bottom, so each successive node up represents an earlier and earlier common ancestor.

In this style,

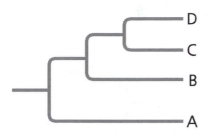

the extant (living) species or groups are listed at the right, so each successive node to the left represents an earlier and earlier common ancestor.

And in this style, time is depicted as concentric

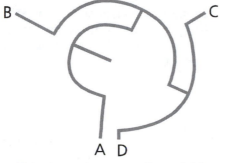

circles. Extant species are on the outside, and earlier and earlier common ancestors can be found closer and closer to the center.

Tips and Relatedness
The tips of phylogenetic trees aren't locked in to where they appear on the page. Relatedness comes from the shared common ancestors—i.e., the nodes. Think of the tree as a mobile, with each group of branches spinning freely at the nodes. This tree

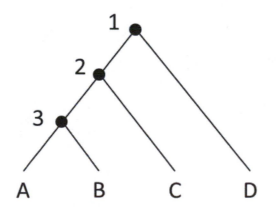

shows exactly the same relationships as this tree:

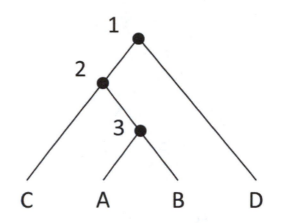

So is C more closely related to D or to A? Node 1 represents the hypothetical common ancestor of all four groups (A-D). Node 2 represents the hypothetical common ancestor of groups A–C, and node 3 represents the hypothetical common ancestor of A and B. C shares a more recent common ancestor with A (node 2) than it does with D (node 1), so C is more closely related to A.

Counting the number of nodes does not indicate relatedness either. C is not more closely related to A because only one node separates them. Again, think of the nodes as points that can swing freely.

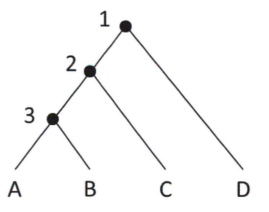

Here, only one node separates C from A and C from D.

And here,

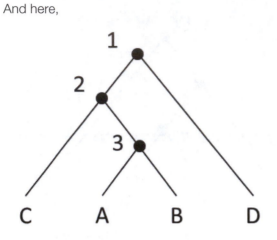

two nodes separate C from A. But C is more closely related to A because it shares a more recent common ancestor with A than it does with D.

Straight Lines and Change

A straight line doesn't mean that a species has not changed—the lines are an artifact of the graphic representation. Take this tree:

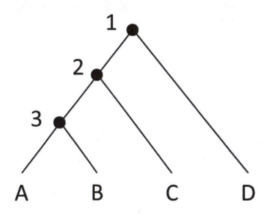

Just because the line to A is the straight line with other lines branching off it, doesn't mean that A didn't change and B–D did. Because the branches can spin freely, A doesn't have to be depicted as a straight line:

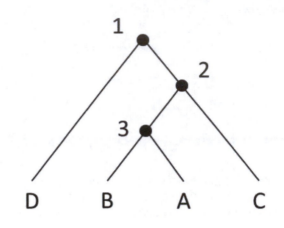

The important thing to understand is that nodes represent the hypothetical common ancestors of the branches, and common ancestors are found earlier and earlier in time.

"Lower" and "Primitive"

Lineages that branch off earlier in the tree are not "lower" or "primitive" forms (see Box 4.1). So D is not more primitive (or higher or advanced) than A. D simply shares an earlier common ancestor with A than B or C.

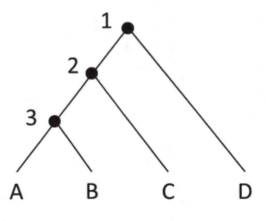

Contemplate

What questions might you ask about fashion using a phylogenetic analysis?	Can technology, such as the evolution of video games, be examined using phylogenies?

Delve Deeper

1. How is a clade depicted in a phylogenetic tree?

2. Is *Tiktaalik* a missing link in the evolution of tetrapods?

3. Why does discovery of the fossil *Liaoconodon* support the hypothesis that separate middle ear bones evolved in two distinct lineages of mammals?

4. If *Homo floresiensis* is a distinct species (and not just a diminutive *Homo sapiens*), then using the phylogeny in Figure 4.18, why would it be considered a distant cousin to our species rather than an ancestor to our species?

5. How does including fossils in phylogenies of extant (living) taxa affect the conclusions scientists can draw?

Go the Distance: Examine the Primary Literature

Bradley Livezey and Richard Zusi developed a phylogeny of modern birds to help resolve some of the conflicting evidence for how birds are related to each other—a heated topic among birders around the world. Many questions still remain, but for now, the birding industry has some restructuring to do in bird books.

- Why did Livezey and Zusi argue that avian systematics needed to be examined?

- Scientists use "outgroups" to determine what characters are "primitive" in the analysis; the organisms being classified are considered the "ingroup." Why did Livezey and Zusi use the Theropoda as an outgroup?

Livezey, B. C., and R. L. Zusi. 2007. Higher-Order Phylogeny of Modern Birds (Theropoda, Aves: Neornithes) Based on Comparative Anatomy. II. Analysis and Discussion. *Zoological Journal of the Linnean Society* 149 (1): 1–95. doi:10.1111/j.1096-3642.2006.00293.x. http://onlinelibrary.wiley.com.weblib.lib.umt.edu:8080/doi/10.1111/j.1096-3642.2006.00293.x/full or http://www.ncbi.nlm.nih.gov/pmc/articles/PMC2517308/.

Test Yourself

1. Which of the following depicts the correct way to read time when examining a phylogeny?

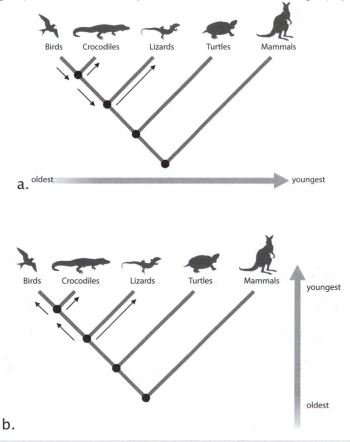

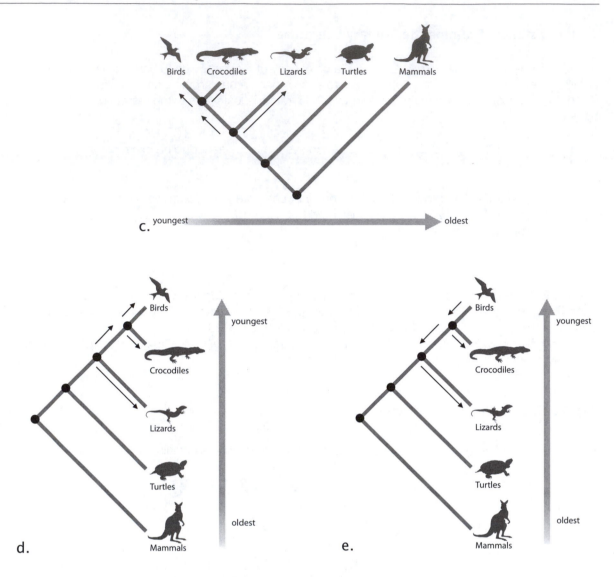

c.

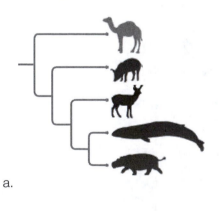

(Adapted from Meir et al. 2007.)

2. Which of the following phylogenies does not indicate the same relationship among whales and other groups?

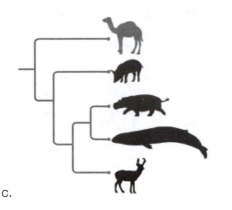

c.

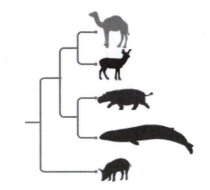

d.

e. All the above show the same relationship.

f. Each phylogeny shows a different relationship.

3. Which of the following statements are depicted by this phylogeny?

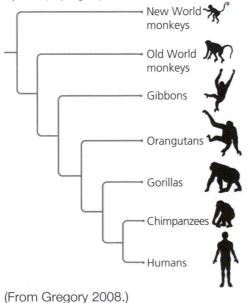

(From Gregory 2008.)

a. The ancestors of humans became gradually more "human-like" over time.

b. Old World monkeys share a common ancestor with humans.

c. Humans represent the end of a lineage of animals whose common ancestor was primate-like.

d. Humans evolved from chimpanzees.

e. None of the above is depicted by this phylogeny.

f. All of the above are depicted by this phylogeny.

4. In the past, morphological evidence alone was used to examine the relationships of the cetaceans to other groups. According to this phylogeny:

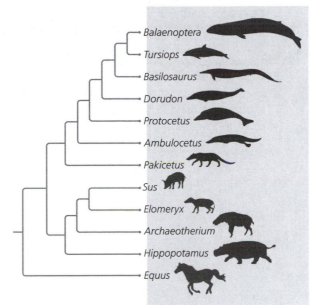

(Modified from Gatesy and O'Leary 2001.)

a. The extinct genus *Ambulocetus* is more distantly related to the genus *Balaenoptera* than to the extinct genus *Protocetus*.

b. The genus *Equus* (horses) is more closely related to the genus *Hippopotamus* (hippos) than to the extinct genus *Pakicetus*.

c. More changes have occurred in the whale lineage than in the horse lineage.

d. Horses (genus *Equus*) were the ancestor of whales and hippos.

e. All of the above are correct statements.

f. None of the above is a correct statement.

5. What homologies do *Tiktaalik* and *Acanthostega* share?
   a. The number of digits at the end of their wrists.
   b. Weight-bearing elbows.
   c. Lungs.
   d. Both a and b.
   e. Both b and c.
   f. All of the above.

6. Which group of theropod dinosaurs did not have feathers according to the evogram in Figure 4.15?
   a. Allosaurids.
   b. Compsognathids.
   c. Tyrannosauroids.
   d. Oviraptorosaurs.
   e. None of the theropod dinosaurs had feathers.

7. What evidence supports the evolution of bipedality in very early hominins?
   a. Anchors for muscles on the pelvis similar to modern humans but unlike those of modern chimpanzees and other apes.
   b. A foramen magnum that is oriented downward similar to modern humans instead of backward like modern chimpanzees and other apes.
   c. Femur bones adapted for supporting an upright torso like in modern humans.
   d. Small stiff toes more similar to modern humans than to those of modern chimpanzees and other apes.
   e. All of the above.

8. What is a phylogeny?
   a. A graphic depiction of the hypothetical relationship among species, populations, or even genes.
   b. A graphic depiction that shows how natural selection caused the evolution of lineages.
   c. A graphic depiction of the morphological similarities among organisms.
   d. All of the above.
   e. None of the above.

Answers for Chapter 4
Check Your Understanding

1. How does homology relate to the theory of evolution?

   a. Homology refers to traits that superficially look alike, like bat wings and human arms; the theory of evolution cannot explain the differences.

   Incorrect. Homology refers to traits that are similar among species because they were inherited from a common ancestor with those traits. Similarities may be "beneath the surface," like bat wings and human arms—the underlying structure is similar in the different species despite overt differences in appearance. Darwin, Cuvier, and many scientists before them (Chapter 2), recognized homology as a basis for organization for comparative biology—that related organisms often had similar morphologies. The theory of evolution provides a mechanism for those observations—descent with modification (see Chapter 2).

   b. Homology refers to traits that look alike but have entirely different origins, like the fins of dolphins and fish; the theory of evolution cannot explain the origins.

   Incorrect. Homology refers to traits that are similar among species because they were inherited from a common ancestor with those traits. Traits that look alike but have different origins are said to be analogous. Darwin, Cuvier, and many scientists before them (Chapter 2), recognized homology as a basis for organization for comparative biology—that related organisms often had similar morphologies. The theory of evolution provides a mechanism for those observations—descent with modification (see Chapter 2).

   c. Homology refers to traits that are structurally similar in different organisms, like bat wings and human arms, because they each were inherited from a shared common ancestor with those traits; the theory of evolution provides a mechanism for those observations.

   Correct. Darwin, and many scientists before him, recognized homology as a basis for organization for comparative biology—that related organisms often had similar morphologies. The fact that many homologies are found together in the same groups of species strengthens support for Darwin's idea of descent with modification (see Chapter 2).

   d. Homology refers to traits that have converged on a shared form; the theory of evolution provides a mechanism for those observations.

   Incorrect. Homology refers to traits that are similar among species because they were inherited from a common ancestor with those traits—like bat wings and human arms—where the underlying structure is similar in the different species. Analogous traits are traits that have converged on a shared form, like the fins of dolphins and fish. Darwin, Cuvier, and many scientists before them recognized homology as a basis for organization for comparative biology—that related organisms often had similar morphologies, and the theory of evolution provides a mechanism for those observations—descent with modification (see Chapter 2).

2. How do scientists know if their explanations are correct?

   a. They don't always know; they can only rely on the quality and quantity of evidence available to them.

   Correct. Scientists often work with the unknown—evidence that is not directly observable. They use experiments and observations to generate explanations that are then tested further. The better an explanation does at predicting new evidence, the more confidence scientists have in it. If a prediction fails, the explanation is scrutinized and altered and tested again. And scientists often test explanations of other scientists that are published in scientific journals with their own experiments to see if they get the same results. So scientists rely on the quality and quantity of evidence available to them to determine the correctness of explanations. New evidence can always alter that determination (see Chapter 2).

   b. They do an experiment, and if it turns out the way they predicted it would, they know they are correct.

   Incorrect. Although scientists use experiments to test their predictions, they rarely accept the outcome of a single experiment as proof of correctness of an explanation. When the outcome of an experiment supports their predictions, scientists consider that evidence for their explanation. The more support for an explanation, the more confidence scientists have in it. Sometimes the outcomes of experiments can be misleading, so scientists often test explanations of other scientists that are published in scientific journals with their own experiments to see if they get the same result. This is an essential part of science, because even an honest scientist can get a misleading result from an experiment. Scientists rely on the quality and quantity of evidence available to them to determine the correctness of explanations. New evidence can always alter that determination (see Chapter 2).

c. They publish the results of their experiments to prove they are correct.

Incorrect. Publication does not necessarily prove the correctness of explanations. Scientists publish the results of their experiments to showcase their findings and provide support for their explanations. And scientists often use the results of other scientists that are published in scientific journals to test explanations with their own experiments to see if they get the same result. The better job that an explanation makes at predicting the new evidence, the more confidence scientists have in it. So scientists rely on the quality and quantity of evidence available to them to determine the correctness of explanations. Additional evidence can always alter that determination (see Chapter 2).

d. Unless scientists can directly observe the phenomenon they are interested in, they cannot know whether their explanations are correct.

Incorrect. Rarely can scientists directly observe the phenomena they are interested in, and even direct observations are not foolproof. Scientists develop explanations that can account for indirect clues gathered through experiment and observation. The better an explanation does at predicting new evidence, the more confidence scientists have in it. If a prediction fails, the explanation is scrutinized and altered and tested again. And scientists often test explanations of other scientists that are published in scientific journals with their own experiments to see if they get the same result. So scientists rely on the quality and quantity of evidence available to them to determine the correctness of

explanations. New evidence can always alter that determination (see Chapter 2).

3. How accurate is radiometric dating?

a. Not very accurate because scientists cannot know how much of a particular isotope was originally present in a rock, and thus they cannot know how much of it has decayed.

Incorrect. It's true that scientists were not there when the rock formed and cannot know how much of a particular isotope was originally present, but they can examine different minerals within a rock and compare the ratios of the different isotopes using fairly straightforward mathematical formulas to determine the rock's age (see Box 3.1).

b. Very accurate because scientists can determine the exact age of rocks and the fossils found within them.

Incorrect. Although radiometric dating is a valuable tool, it cannot be used to determine the exact ages of rocks. Radiometric dating provides estimates of age based on the half-life of the isotopes within the rocks. Different isotopes can be used to measure different time scales, and the time scale being addressed can affect the precision of the estimate. Isotopes with very long half-lives can be used to determine the ages of very old rocks with relative accuracy—say a 3-million-year range for stromatolite fossils dating back 3.4 billion years. Isotopes with shorter half-lives can provide more precise estimates for younger rocks—for example, a margin of error of 7000 years (see Box 3.1).

c. Accurate enough because radiometric dating can determine estimates for the ages of rocks and fossils, often with relatively small margins of error.

Correct. Radiometric dating provides estimates of age based on the half-life of the isotopes within the rocks. Different isotopes can be used to measure different time scales, and the time scale being addressed can affect the precision of the estimate. Isotopes with very long half-lives can be used to determine the ages of very old rocks with relative accuracy—say a 3-million-year range for stromatolite fossils dating back 3.4 billion years. Isotopes with shorter half-lives can provide more precise estimates for younger rocks—for example, a margin of error of 7000 years. Scientists can also use different elements to check and hone their estimates (see Box 3.1).

d. Not very accurate because scientists use probabilities to determine decay rates for each isotope they use in radiometric dating. Incorrect. Scientists definitely use probabilities to determine decay rates, but the use of statistical techniques does not negate the accuracy of radiometric dating as a methodology. Statistics are validated and scrutinized techniques that can provide insight to the accuracy and precision of measures, such as the age of rocks. Scientists accept that no measure is completely accurate and often use several lines of evidence to improve their estimates. See Box 3.1 for a discussion of the role of different isotopes in estimating the ages of rocks.

Identify Key Terms

branch	A lineage evolving through time between successive speciation events.
clade	A single "branch" in the tree of life representing an organism and all of its descendants.
common ancestor	The ancestral form or species from which two or more different forms or species descended.
homology	The similarity of characteristics in different species because of a common ancestor.
hypothesis	A proposed explanation for an observable phenomenon, usually forming a basis for possible experiments to confirm its viability.
node	A point in a phylogeny where a lineage splits (a speciation event).
phylogeny	A visual representation of the evolutionary history of populations, genes, or species.
tetrapod	A member of the superclass Tetrapoda that includes the first four-limbed vertebrates and their descendants, including amphibians, birds, and mammals.
therian	A member of the subclass Theria within the class Mammalia that includes marsupial and placental mammals.
tip	The terminal end of an evolutionary tree, representing the species, the molecule, or the population that is to be compared to other species, molecules, or populations.

Link Concepts

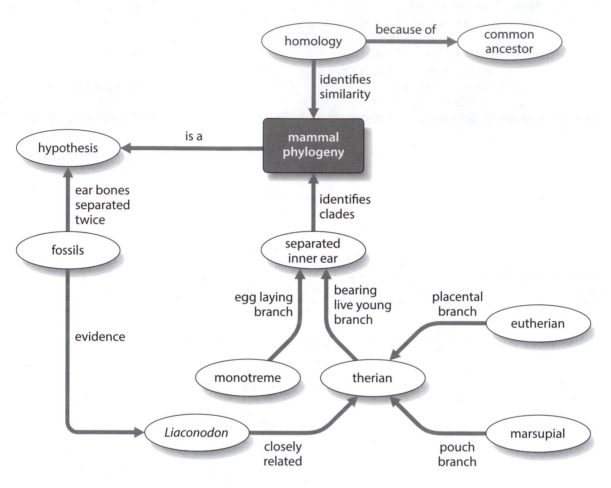

Games and Exercises

	Vomer present Cerci unsegmented	Thorn pads on male T10 Fore femora curved	Abdominal sterna Cerci dorsoventrally flat	Loss of tergal thorn pads
Timemetodea	Yes	No	No	No
Euphasmatodeans	Yes	Yes	No	No
Nanophyllium	Yes	Yes	Yes	No
Microphyllium	Yes	Yes	Yes	Yes
Chitoniscus	Yes	Yes	Yes	Yes
Phyllium	Yes	Yes	Yes	Yes

- Are the Euphasmatodea a monophyletic clade?
  Yes, the Euphasmatodea are depicted as a monophyletic clade—the branch consists of the common ancestor and all the descendent taxa. However, the cladogram indicates that there is no clear resolution of the extant taxa within the Euphasmatodea.

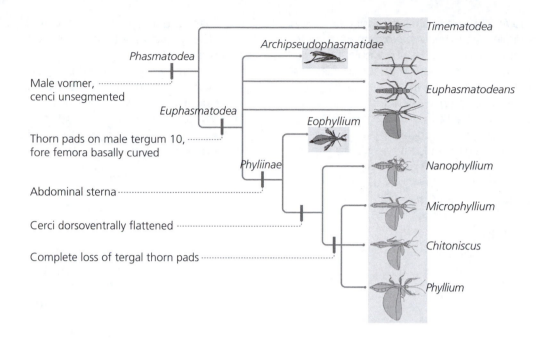

- Did the addition of *Eophyllium* clarify the resolution of the Euphasmatodea?

  No. *Eophyllium* has abdominal sterna, so it is considered a member of the Phylliinae, leaf insects that exhibit an extreme form of morphological and behavioral leaf mimicry. Since it doesn't have dorsoventrally flattened cerci though, Eophyllium must have arisen before the *Nanophyllium, Microphyllium, Chitoniscus*, and *Phyllium* clade split off. Archipseudophasmatidae, however, must be a member of the Euphasmatodea because it lacks abdominal sterna. However, this extinct group provides little resolution for the extant Euphasmatodea.

Delve Deeper

1. How is a clade depicted in a phylogenetic tree?

   A clade is an organism and all of the descendants of that organism, including any new lineages. It is a single branch of an evolutionary tree. A phylogenetic tree represents many different clades, and defining a clade depends on the level of interest. A specific clade starts at a node and includes all the branches and nodes below it, and clades can be nested within clades.

2. Is *Tiktaalik* a missing link in the evolution of tetrapods?

   No. "Missing link" is a misnomer for a direct ancestor. Not only are scientists NOT looking for missing links, they also do not expect to find them because of the immense diversity of historic life and the extremely low probability of fossilization. *Tiktaalik* definitely resolves some of the questions about the evolution of the tetrapod clade though: the common ancestor of tetrapods and their closest living relatives had stout, paddle-shaped fins and a neck.

3. Why does discovery of the fossil *Liaoconodon* support the hypothesis that separate middle ear bones evolved in two distinct lineages of mammals?

   Liaoconodon is an extinct species of mammal. The age of the fossil indicates that it is older than extant monotremes and therians (see Figure 4.13), but its morphology indicates that it is more similar to therians than to monotremes. Both extant therians and extant monotremes have middle ear bones that are separated from the jaws in the adults, but the bones homologous to therian middle ear bones in *Liaoconodon* are still connected to the jaws. That means that during the evolution of mammals the middle ear bones had to have separated two different times, once in the monotreme lineage and once in the therian lineage after modern therians split off from *Liaoconodon*.

4. If *Homo floresiensis* is a distinct species (and not just a diminutive *Homo sapiens*), then using the phylogeny in Figure 4.18, why would it be considered a distant cousin to our species rather than an ancestor to our species?

Because *H. floresiensis* is related to *H. erectus* and on a separate branch of the clade. "Cousins" describes species separated by more than one node, and at least two nodes separate *H. erectus* from *H. sapiens*. We share a common ancestor with *H. erectus*, but *H. erectus* is not our ancestor.

Test Yourself

1. b; 2. d; 3. b; 4. f; 5. e; 6. a; 7. e; 8. a

5. How does including fossils in phylogenies of extant (living) taxa affect the conclusions scientists can draw?

Including fossils can change the hypothesis generated by the phylogeny. Fossils can define the timing of branching events. They can affect understanding of common ancestors. The discovery of new fossils can also generate new questions about clades.

Chapter 5
EVOLUTION'S RAW MATERIALS

Check Your Understanding

A

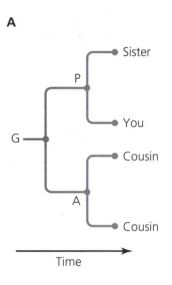

B

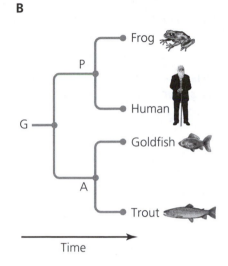

1. How can the phylogeny on the left be used to understand the phylogeny on the right?
   a. The phylogeny on the left shows that you and your sister are more closely related to each other than you are to your cousins, just as humans and frogs are more closely related to each other than they are to goldfish and trout.
   b. The phylogeny on the left shows that the relationship between humans and frogs cannot be compared with the relationship between siblings.
   c. The phylogeny on the left shows that your cousins must be more closely related to each other than you and your sister because trout and goldfish are more closely related to each other than frogs and humans.
   d. The phylogeny on the right shows that humans are more closely related to goldfish than to trout.

2. Who incorporated the idea that the theory of evolution required the capacity for one generation to pass on its traits to the next?
   a. Charles Darwin.
   b. Jean-Baptiste Lamarck.
   c. Alfred Russel Wallace.
   d. All of the above.
   e. None of the above.

3. Why are mutations important in evolution?
   a. Because mutations are always deleterious, and organisms with these deleterious mutations do not survive and reproduce.
   b. Because if mutations are random, then life must have evolved by chance.
   c. Because evolution is the process of mutated genes becoming more or less common over the course of generations.
   d. Because the environment causes mutations that organisms can use.

Learning Objectives for Chapter 5:

➢ Describe the molecular architecture of proteins.
➢ Describe the structural organization of DNA within a human cell.
➢ Compare and contrast the events that occur in transcription and translation.
➢ Explain the difference between coding and noncoding segments of DNA.
➢ Explain why genetic recombination is important to evolution.
➢ Explain how variation in a phenotype can range around a mean.
➢ Explain why most phenotypes cannot be directly linked to genotypes.

Identify Key Terms

Connect the following terms with their definitions on the right:

allele	The process that takes place when RNA polymerase reads a coding sequence of DNA and produces a complementary strand of RNA, called messenger RNA (mRNA).
chromosome	A region of DNA that initiates transcription of a particular gene, usually found upstream on the same strand.
codon	Any process in which genetic material is transferred to another organism without descent.
DNA	The process by which information from a gene is transformed into a product.
dominant allele	An essential macromolecule for life assembled from nucleotides that link together to form a structural backbone. Nucleotides include the sugar (ribose) and one of four bases: adenine (A), cytosine (C), guanine (G), and uracil (U).
gene	One of any number of alternative forms of a gene or genetic locus.
gene expression	The process that takes place when a strand of messenger RNA (mRNA) is decoded by a ribosome to produce a strand of amino acids.
genetic recombination	An observable, measurable characteristic of an organism, such as a morphological structure (e.g., antlers, muscles), a developmental process (e.g., learning), a physiological process or performance trait (e.g., running speed), a behavior (e.g., mating display), or even the molecules produced by genes (e.g., hemoglobin).
genotype	A DNA sequence that resembles functional genes but has lost its protein-coding ability or

	is no longer expressed. These sequences often form after a gene has been duplicated, when one, or more, of the redundant copies subsequently loses its function.
heterozygote	The crossover and exchange of genetic material between paired chromosomes during meiosis that can lead to new combinations of alleles. This process is an important source of heritable variation.
homozygote	The genetic makeup of an individual, including all the alleles of all the genes in that individual. The term is often used to refer to the specific alleles carried by an individual for any particular gene.
horizontal gene transfer	A group of RNAs that act to regulate gene expression. These noncoding molecules bind to complementary sequences on specific messenger RNAs (mRNAs) and can enhance or silence the translation of genes. The human genome encodes more than 1000 of these tiny RNAs.
messenger RNA (mRNA)	Parasite-like segments of DNA that only make new copies of themselves that are reinserted in different places in the genome, such as transposons ("jumping genes") and plasmids.
microRNA	An allele visible (evident in the phenotype) in a heterozygote. Complete dominance occurs when the phenotype of the heterozygote is identical to the homozygote.
mobile genetic element	An organism that carries two copies of the same allele (such as rr or RR).
mutation	An essential macromolecule for life formed from chains of building blocks known as amino acids. These molecules can form complex shapes or join with other such molecules to form even larger structures. They give cells structure, break down other molecules, store important molecules, and deliver information, relaying signals within a cell or between cells.
phenotype	An organism that carries two different alleles of a gene (such as Rr).
promoter region	Any change to the genomic sequence of an organism.
protein	A sequence of three nucleotides that designates an amino acid as part of genetic code for protein synthesis.

pseudogene		An essential macromolecule for life assembled from nucleotides that link together to form a structural backbone. Nucleotides include the sugar (deoxyribose) and one of four bases: adenine (A), cytosine (C), guanine (G), and thymine (T).
recessive allele		An RNA molecule that is transcribed from a gene and conveys the information necessary to assemble a protein within a ribosome. These molecules specify sequences of amino acids, three bases at a time, each used to identify one of 20 amino acids.
RNA		A tightly bundled rod of DNA found in eukaryotes. Humans have two types: 44 that can be arranged into identical pairs, and two that pair, but differ between males and females.
transcription		Segments of DNA whose nucleotide sequences code for proteins, or RNA, or regulate the expression of other genes.
translation		An allele that usually has little or no effect on the phenotype when it is expressed in heterozygous individuals.

Link Concepts

1. Turning the information coded in DNA into proteins is an amazing process. A messenger RNA (mRNA) molecule is transcribed from a gene, and a ribosome uses that mRNA to assemble a protein. The ribosome reads three bases (a codon) in the mRNA at a time, each matching one of 20 amino acids. Use the chart below to transcribe the two strands of DNA to mRNA and add the amino acids by codon.

Strand 1:   CACGTGGACTGAGGACTC

Strand 2:   CACGTGGACTGAGGACAC

- What's the difference between the two strands?

		2nd base in Codon				
		U	C	A	G	
1st base in Codon	U	Phe Phe Leu Leu	Ser Ser Ser Ser	Tyr Tyr STOP STOP	Cys Cys STOP Trp	U C A G
	C	Leu Leu Leu Leu	Pro Pro Pro Pro	His His Gln Gln	Arg Arg Arg Arg	U C A G
	A	Ile Ile Ile Met	Thr Thr Thr Thr	Asn Asn Lys Lys	Ser Ser Arg Arg	U C A G
	G	Val Val Val Val	Ala Ala Ala Ala	Asp Asp Glu Glu	Gly Gly Gly Gly	U C A G

- What's the difference between the amino acid sequences that result?

- If the mutation in Strand 2 was at the third base in that codon instead of the second, would the hemoglobin protein form normally?

HBB is the gene associated with sickle-cell anemia. Scientists have pinpointed the location of this gene on Chromosome 11 in humans. They have determined that 444 base pairs within the mRNA code for the amino acid sequence of the gene's protein product, hemoglobin, and that protein itself is 146 amino acids long (the mRNA includes 2 extra base pairs). Several hundred variations of the HBB gene are known, but sickle-cell anemia is most commonly caused by the hemoglobin variant Hb S. This variant made in the bodies of people with sickle-cell disease (Hb S) differs from normal hemoglobin (Hb A) in just one amino acid of the HBB polypeptide chain. In normal hemoglobin, this amino acid is glutamic acid. In sickle-cell hemoglobin, it is a valine.

2. Develop your own concept map that illustrates the relationships between hemoglobin alleles and sickle-cell anemia. Think about concepts such as mutation, allele, amino acid, base, DNA, heredity, and phenotype.

Interpret the Data

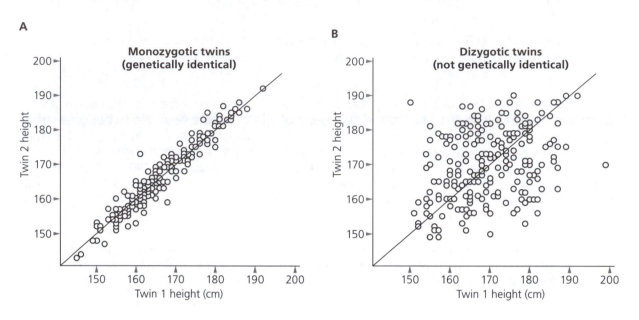

Figures 5.14 A and B show the relationship between the heights of identical and fraternal twins. One twin's height is marked along the x-axis, and his or her twin sibling's height is marked on the y-axis. (Data courtesy of David Duffy.)

In part A of the graph, can you predict the height of Twin 2 when twins are identical? For example, what would you expect the height of Twin 2 to be if the height of Twin 1 is 160 cm?

Is there variation in the heights of identical twins? Do heights match up exactly?

In part B of the graph, can you accurately predict the height of Twin 2 if fraternal Twin 1 is 160 cm?

Is there any correlation in height between fraternal twins?

What might be contributing to the variation between fraternal twins?

Games and Exercises

Smiley Face Trait Inheritance

Here's a fun way to think about traits and inheritance adapted from T. Tomm (2003) and http://sciencespot.net/.

You'll need a nickel and a dime—the nickel will represent the female parent and the dime the male parent. Both parents are heterozygous for all the Smiley Face traits. Flip both coins for each trait in the table below. If the coin lands with heads up, that parent contributes a dominant allele for that trait. If the coin lands tails up, that parent contributes a recessive allele. Circle the allele the offspring receives from each parent for each trait by circling the corresponding letter. Use the results and the Smiley Face Traits chart below to fill in the genotype and phenotype for each trait. Then, create a sketch of your smiley face.

Trait	Female		Male		Genotype	Phenotype
Face Shape	C	c	C	c		
Eye Shape	E	e	E	e		
Hair Style	S	s	S	s		
Smile	T	t	T	t		
Ear Style	V	v	V	v		
Nose Style	D	d	D	d		
Face Color	Y	y	Y	y		
Eye Color	B	b	B	b		
Hair Length	L	l	L	l		
Freckles	F	f	F	f		
Nose Color	R	Y	R	Y		
Ear Color	P	T	P	T		

Face Shape	Nose Style	
Circle (C) Oval (c)	Down (D) Up (d)	
Eye Shape Star (E) Blast (e)	**Face Color** Yellow (Y) Green (y)	**Face Color** Blue (B) Red (b)
Hair Style Straight (S) Curly (s)	**Hair Length** Long (L) Short (l)	**Hair Length** Blue (B) Red (b)
Smile Thick (T) Thin (t)	**Nose Color** Red (RR) Orange (RY) Yellow (YY)	**Ear Color** Hot Pink (PP) Purple (PT) Teal (TT)
Ear Style Curved (V) Pointed (v)	**Sex** To determine the sex, flip the coin for the male parent. Heads equals X and tails equals Y. XX – Female – Add pink bow in hair XY – Female – Add blue bow in hair	

Extract Your Own DNA

Anna Rothschild produced this short video for NOVA showing how you can extract your own DNA using bottled water, clear dish soap, food coloring, table salt, and 70 percent isopropyl alcohol.

http://www.pbs.org/wgbh/nova/body/extract-your-dna.html

Go Online

WEHI.TV, of the Walter and Eliza Hall Institute, produces awesome animations that illustrate scientific concepts that are difficult to observe. The following links illustrate transcription and translation.

http://www.wehi.edu.au/education/wehitv/dna_central_dogma_part_1_-_transcription/

http://www.wehi.edu.au/education/wehitv/dna_central_dogma_part_2_-_translation/

The entire 8-minute video by Drew Berry explaining DNA transcription using amino acids to turn genes into proteins can be found here on YouTube.
http://www.youtube.com/watch?v=TSv-Rq5C3K8

Learn.Genetics, produced by the University of Utah, is an incredible resource for learning about genetics and its role in evolution. Below are two modules that are particularly relevant.

TOUR OF THE BASICS

This module provides a narrated overview of the basics—what is DNA, what is a gene, what is a chromosome, what is a protein, what is heredity, and what is a trait?

http://learn.genetics.utah.edu/content/begin/tour/

TRANSCRIBE AND TRANSLATE A GENE

This module is a cool interactive that lets you transcribe a gene and build a protein.

http://learn.genetics.utah.edu/content/begin/dna/transcribe/

Gene Expression

From genomicseducation, this YouTube video is an animation that illustrates gene expression within a cell.

http://www.youtube.com/watch?v=OEWOZS_JTgk

Signal Transmission and Gene Expression

Paul Anderson, of bozemanbiology, takes the animation one step further and shows what extreme base jumping can teach about gene expression and the production of glucose within the cells of the liver

http://www.youtube.com/watch?v=D-usAds_-lU

The Gene School Interactive page includes a number of different experiments and games to play that all have to do with genetics.

http://library.thinkquest.org/19037/teach_links.html

The Genetic Basis for Bacterial Mercury Methylation

Jerry M. Parks and colleagues identify two genes involved in the mercury methylation by bacteria in a recent issue of Science magazine.

http://www.sciencemag.org/content/339/6125/1332.abstract

Patients Should Get DNA Information, Report Recommends

In *Science* Insider, the American College of Medical Genetics and Genomics recently argued that patients whose genomes are sequenced for medical reasons should be given information about genes that may affect their health, including those that put them at risk of certain cancers and potentially fatal heart conditions, even if those risks are unrelated to the reason for sequencing in the first place.

http://news.sciencemag.org/health/2013/03/patients-should-get-dna-information-report-recommends?ref=em

Common Misconceptions

Our understanding of genetics has changed significantly since Mendel's time, and today it continues to richen and deepen. Some of the models we use to understand the complexities of genetics can sometimes lead to misconceptions.

Rarely Are Traits Influenced by a Single Locus
Mendel discovered that some phenotypes were predictable, and some diseases, such as Huntington's, can be tied to a single mutation. The idea that a single gene "codes for" a specific phenotypic effect is often an oversimplification, however. The majority of traits are the result of polygenic effects (many genes).

Genes Are Not Analogous to Words in a Sentence
Thinking of genes as words in a sentence is another oversimplification. Genes are actually quite complex concepts, and the more scientists learn, the less acceptable this oversimplification becomes. For example, DNA sequences for proteins include exons that carry instructions and noncoding introns, both of which may mix and match and code for different proteins—protein-encoding regions may not have borders.

A Locus Can Have More Than Two Alleles
Simple models of loci (plural of locus) often represent the alleles in upper- and lowercase letters, such as AA, Aa, and aa, implying that the locus only has two alleles. A more appropriate way of representing alleles incorporates subscripts, such as A_1, A_2, A_3, A_4, reflecting the diverse number of alleles that may be known for any locus.

Phenotypes Are Influenced by More than Just Genotypes

Genes are clearly an important influence on how an organism turns out, but environmental conditions are also important. When scientists refer to environmental conditions, they consider more than the external environment (e.g., light conditions). A host of internal environmental factors influence gene expression, including other gene products, such as the timing and quantity of the release of the hormone adrenaline from the adrenal gland. This hormone circulates around the body and attaches to the surface of muscles and other types of cells, switching on other genes inside.

Contemplate

Why are mutations important to evolution by natural selection?	If alcoholism is heritable, what factors can influence whether the children of an alcoholic will be alcoholics?

Delve Deeper

1. What are the possible outcomes to a protein if the gene that codes for it has a point mutation?

2. Explain how a gene in one cell can influence the development of a different cell through the action of a hormone.

3. What are two ways sexual reproduction can lead to offspring that are genetically different from their parents?

4. Why aren't the mutations that occur in skin cells or in other organs, such as the heart or brain, heritable?

5. Why don't all phenotypic traits occur as discrete, alternative states like Mendel's peas?

Go the Distance: Examine the Primary Literature

As Chapter 5 explains, height is a classic polygenic trait that is highly heritable. Hana Lango Allen and colleagues examined the genetic variation of 183,727 different people and found that at least 180 loci influence adult height, but they were only able to explain 10 percent of the variation. Nevertheless, their methods offer insight to the complexities of the genetic architecture of these kinds of traits.

- Why do they argue their approach is so valuable?

• How might this approach help our understanding of polygenic diseases?

Allen, H. L., K. Estrada, G. Lettre, S. I. Berndt, M. N. Weedon, et al. 2010. Hundreds of Variants Clustered in Genomic Loci and Biological Pathways Affect Human Height. *Nature* 467: 832–38. doi:10.1038/nature09410.

Test Yourself

1. What is the name of DNA sequences that have lost their protein-coding ability?
   a. Transposons.
   b. Pseudogenes.
   c. Introns.
   d. Both a and b.
   e. Both b and c.

2. When an organism incorporates genetic material from another organism that is not its parent, this is known as:
   a. Vertical gene transfer.
   b. Diagonal gene transfer.
   c. Horizontal gene transfer.
   d. None of the above.

3. What is an allele?
   a. One of several alternative forms of the DNA sequence of the same locus.
   b. One of several alternative forms of a gene that occur at the same place on paired chromosomes.
   c. One of several alternative forms of a gene that occur at the same locus in different individuals.
   d. All of the above.
   e. None of the above.

4. What is the most important factor generating genetic diversity in eukaryotes?
   a. Horizontal gene transfer.
   b. Mutation.
   c. Genetic recombination.
   d. Both a and b.
   e. Both b and c.

5. What is an organism's phenotype?
   a. The interaction of an organism's genes with the environment to produce characteristics, such as the effect of food to produce horns.
   b. Characteristics of an organism that can be classified into discrete categories, such as gender or eye color.
   c. Any aspect of an organism that can be measured, such as how it looks, how it behaves, how it's structured.
   d. The genetic makeup of an individual.
   e. Both b and d.

6. Does a genotype always produce the same phenotype?
   a. No. A genotype can produce multiple phenotypes because individuals have different alleles—some dominant and some recessive—the combination of which results in different phenotypes.
   b. No. Some genotypes can produce multiple phenotypes depending on the internal and external environments.
   c. No. Some genotypes can produce multiple phenotypes but only when the amount of nutrition available varies.
   d. Yes. Genotypes are very specific, defining everything from eye color to height.
   e. Yes. A single genotype is directly linked to the phenotype through DNA.

7. Which type of mutation best describes the type of mutation when one base pair in a sequence is changed into another?
   a. Point mutation.
   b. Insertion.
   c. Deletion.
   d. Duplication.
   e. Inversion.

8. How can mutations add to the variation within a population?
   a. They can alter the phenotype of some individuals by affecting the production of proteins such as enzymes, signals, receptors, or structural elements; for example, they may

cause some individuals to be able to carry more oxygen in their blood than other individuals.

b. They can alter the phenotype of some individuals by affecting if and when genes are expressed, so that some individuals may have longer horns than other individuals.

c. They alter the phenotypes of some individuals by affecting the production of proteins such as enzymes, signals, receptors, or structural elements, but because these mutations are always harmful they reduce the variation within a population.

d. They alter the phenotypes of some individuals by affecting the production of proteins such as enzymes, signals, receptors, or structural elements, but because these mutations never occur in eggs or sperm, they do not affect the variation within a population.

e. Both a and b.

f. Both c and d.

g. All of the above.

Answers for Chapter 5
Check Your Understanding

1. How can the phylogeny on the left be used to understand the phylogeny on the right?

A

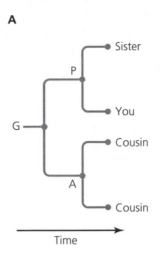

Time

B

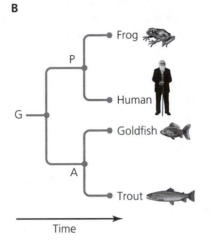

Time

a. The phylogeny on the left shows that you and your sister are more closely related to each other than you are to your cousins, just as humans and frogs are more closely related to each other than they are to goldfish and trout.
Correct. Although the phylogeny on the left shows the relationships among individual family members and not groups of organisms, the idea of common ancestry is essentially the same. Humans are more closely related to frogs than they are to fish because they share a more recent common ancestor with frogs (P)—just like you and your sister share a more recent common ancestor (P) than you and your cousins (G). Humans share a more distant common ancestor with fish (G) (see Chapter 4).

b. The phylogeny on the left shows that the relationship between humans and frogs cannot be compared with the relationship between siblings.
Incorrect. Although the phylogeny on the left shows the relationships among individual family members and not groups of organisms, the idea of common ancestry is essentially the same. You and your sister share a more recent common ancestor (P) than you and your cousins (G). Similarly, humans and frogs share a more recent common ancestor than humans and trout (see Chapter 4).

c. The phylogeny on the left shows that your cousins must be more closely related to each other than you and your sister because trout

and goldfish are more closely related to each other than frogs and humans.
Incorrect. You and your sister share a common ancestor (P), and each of your cousins also shares a common ancestor (A). You share a more distant common ancestor with your cousins (G). The tree on the left indicates that the relationship between you and your sister is similar to the relationship between your cousins. The phylogeny on the right shows that humans and frogs share a common ancestor (P) and, similarly, trout and goldfish share a common ancestor (A). So humans and frogs are more closely related to each other because they share a more recent common ancestor than with goldfish, and trout and goldfish are more closely related to each other because they share a more recent common ancestor than with humans (see Chapter 4).

d. The phylogeny on the right shows that humans are more closely related to goldfish than to trout.
Incorrect. The branches on a phylogeny can swing freely, just like the different parts of a mobile. So the branch with you and your sister can rotate freely on its axis, and the branch with humans and frogs can rotate freely on its axis. If you and your sister were flipped, that wouldn't make her any more related to your cousins than you are. You and she share a more recent common ancestor than you and your cousins. Likewise, if

humans and frogs were flipped, that doesn't make frogs any more related to goldfish (or trout, if those two were flipped). Humans and frogs share a more recent common ancestor than they do with fish (see Chapter 4).

2. Who incorporated the idea that the theory of evolution required the capacity for one generation to pass on its traits to the next?

   a. Charles Darwin.

   Correct, but so are other answers. Darwin conducted exhaustive studies that indicated a capacity for one generation to pass on its traits to the next, but he had no explanation for how that transference took place (see Chapter 2). Although Darwin's evidence was substantial and significant, scientists have discovered much about the mechanisms for passing on traits.

   b. Jean-Baptiste Lamarck.

   Correct, but so are other answers. Lamarck incorporated the capacity for one generation to pass on its traits to the next in his ideas about evolution, but he believed these traits could be acquired in an individual's lifetime. Lamarck's ideas were intuitive at the time, but scientists later rejected his ideas about inheritance of acquired characteristics (see Chapter 2). Since the early 1900s, scientists have discovered much about the mechanisms for passing on traits.

   c. Alfred Russel Wallace.

   Correct, but so are other answers. Wallace developed a mechanism for evolution very similar to Darwin's idea of natural selection (see Chapter 2). That mechanism requires that traits be heritable, passed down from one generation to the next, and that possession of those traits confers some benefit in terms of reproductive success or survival.

   d. All of the above.

   Correct. Each of these naturalists incorporated some capacity for one generation to pass on traits to another (see Chapter 2), although none of them understood the mechanism by which that transference took place. Since the early development of the theory of evolution, scientists have discovered much about the mechanisms for passing on traits.

   e. None of the above.

   Incorrect. Each of these naturalists incorporated some capacity for one generation to pass on traits to another (see Chapter 2), although none of them

understood the mechanism by which that transference took place. Since the early development of the theory of evolution, scientists have discovered much about the mechanisms for passing on traits.

3. Why are mutations important in evolution?

   a. Because mutations are always deleterious, and organisms with these deleterious mutations do not survive and reproduce.

   Incorrect. Mutations can be deleterious, but they can also be beneficial, or even neutral. Beneficial mutations may lead to greater survival or reproduction through their effects on the characteristics, or phenotype, of an organism. But the effect of a mutation may often be influenced by all the other genes an organism carries (see Chapter 1). Most importantly, mutations create the variation among individuals on which the mechanisms of evolution can act.

   b. Because if mutations are random, then life must have evolved by chance.

   Incorrect. Mutations are random—they can be deleterious, beneficial, or even neutral. But the spread or loss of those mutations within populations by natural selection is definitely not random. Beneficial mutations may lead to greater survival or reproduction through their effects on the characteristics, or phenotype, of an organism. Mutations create the random variation among individuals on which the mechanisms of evolution can act. Natural selection leads to the nonrandom spread or loss of mutations over the course of generations (see Chapter 1).

   c. Because evolution is the process of mutated genes becoming more or less common over the course of generations.

   Correct. Mutations are random—they can be deleterious, beneficial, or even neutral. Beneficial mutations may lead to greater survival or reproduction through their effects on the characteristics, or phenotype, of an organism. As a result, natural selection leads to the nonrandom spread of those beneficial mutations over the course of generations (see Chapter 1). So mutations create the variation among individuals on which the mechanisms of evolution can act.

   d. Because the environment causes mutations that organisms can use.

   Incorrect. Factors in the environment may influence the introduction of errors during the replication of DNA, but those factors do not determine whether those mutations are useful to an organism. Mutations are random—they can be beneficial, neutral, or deleterious. Natural selection leads to the nonrandom spread or loss of those mutations over the course of generations (see Chapter 1). The fact that mutations arise is unrelated to how useful those mutations may be, but mutations do create the variation among individuals on which the mechanisms of evolution can act.

Identify Key Terms

allele	One of any number of alternative forms of a gene or genetic locus.
chromosome	A tightly bundled rod of DNA found in eukaryotes. Humans have two types: 44 that can be arranged into identical pairs, and two that pair, but differ between males and females.
codon	A sequence of three nucleotides that designates an amino acid as part of genetic code for protein synthesis.
DNA	An essential macromolecule for life assembled from nucleotides that link together to form a structural backbone. Nucleotides include the sugar (deoxyribose) and one of four bases: adenine (A), cytosine (C), guanine (G), and thymine (T).
dominant allele	An allele visible (evident in the phenotype) in a heterozygote. Complete dominance occurs when the phenotype of the heterozygote is identical to the homozygote.
gene	Segments of DNA whose nucleotide sequences code for proteins, or RNA, or regulate the expression of other genes.
gene expression	The process by which information from a gene is transformed into a product.
genetic recombination	The crossover and exchange of genetic material between paired chromosomes during meiosis that can lead to new combinations of alleles. This process is an important source of heritable variation.
genotype	The genetic makeup of an individual, including all the alleles of all the genes in that individual. The term is often used to refer to the specific alleles carried by an individual for any particular gene.
heterozygote	An organism that carries two different alleles of a gene (such as Rr).
homozygote	An organism that carries two copies of the same allele (such as rr or RR).
horizontal gene transfer	Any process in which genetic material is transferred to another organism without descent.
messenger RNA (mRNA)	An RNA molecule that is transcribed from a gene and conveys the information necessary to assemble a protein within a ribosome. These molecules specify sequences of amino acids, three bases at a time, each used to identify one of 20 amino acids.
microRNA	A group of RNAs that act to regulate gene expression. These noncoding molecules bind to complementary sequences on specific messenger RNAs (mRNAs) and can enhance or silence the translation of genes. The human genome encodes more than 1000 of these tiny RNAs.
mobile genetic element	Parasite-like segments of DNA that only make new copies of themselves that are reinserted in different places in the genome, such as transposons ("jumping genes") and plasmids.
mutation	Any change to the genomic sequence of an organism.

phenotype	An observable, measurable characteristic of an organism, such as a morphological structure (e.g., antlers, muscles), a developmental process (e.g., learning), a physiological process or performance trait (e.g., running speed), a behavior (e.g., mating display), or even the molecules produced by genes (e.g., hemoglobin).
promoter region	A region of DNA that initiates transcription of a particular gene, usually found upstream on the same strand.
protein	An essential macromolecule for life formed from chains of building blocks known as amino acids. These molecules can form complex shapes or join with other such molecules to form even larger structures. They give cells structure, break down other molecules, store important molecules, and deliver information, relaying signals within a cell or between cells.
pseudogene	A DNA sequence that resembles functional genes but has lost its protein-coding ability or is no longer expressed. These sequences often form after a gene has been duplicated, when one, or more, of the redundant copies subsequently loses its function.
recessive allele	An allele that usually has little or no effect on the phenotype when it is expressed in heterozygous individuals.
RNA	An essential macromolecule for life assembled from nucleotides that link together to form a structural backbone. Nucleotides include the sugar (ribose) and one of four bases: adenine (A), cytosine (C), guanine (G), and uracil (U).
transcription	The process that takes place when RNA polymerase reads a coding sequence of DNA and produces a complementary strand of RNA, called messenger RNA (mRNA).
translation	The process that takes place when a strand of messenger RNA (mRNA) is decoded by a ribosome to produce a strand of amino acids.

Link Concepts

Strand 1:   CACGTGGACTGAGGACTC
mRNA: GUGCACCUGACUCCUGAG
Amino acids: Val His Leu Thr Pro Glu

Strand 2:   CACGTGGACTGAGGACAC
mRNA: GUGCACCUGACUCCUGUG
Amino acids: Val His Leu Thr Pro Val

- What's the difference between the two strands?
  The second base in the last codon is an A rather than a T.
- What's the difference between the amino acid sequences that result?
  Val replaces Glu in Strand 2.
- If the mutation in Strand 2 was at the third base in that codon instead of the second, would the hemoglobin protein form normally?
  Only if the mutation resulted in an A in that position. If the mutation resulted in a U or a C at that position, the amino acid in the chain would be Asp not Glu.

Interpret the Data

- In part A of the graph, can you predict the height of Twin 2 when twins are identical? For example, what would you expect the height of Twin 2 to be if the height of Twin 1 is 160 cm?
  Yes. You can predict the height of Twin 2 with some certainty when twins are identical. If Twin 1 is 160 cm, Twin 2 is likely to be about 160 cm. In part A, the height of Twin 2 ranged from approximately 152 to 165 cm when Twin 1 was exactly 160 cm.
- Is there variation in the heights of identical twins? Do heights match up exactly?
  Yes. Heights of identical twins vary—their heights do not match up exactly. Even though the twins share identical genes, other factors, such as subtle differences in the environment within which the individuals develop, affect the height of each individual.
- In part B of the graph, can you accurately predict the height of Twin 2 if fraternal Twin 1 is 160cm?
  No. The heights of Twin 2 ranged from approximately 158 to 179 when Twin 1 was 160 cm, so an accurate prediction is much more difficult.
- Is there any correlation in height between fraternal twins?
  Although predicting the heights of fraternal twins is not as straightforward as predicting the heights of identical twins, the heights of fraternal twins should still be correlated because the individuals are related and they likely experienced generally similar environments as they developed.
- What might be contributing to the variation between fraternal twins?
  Identical twins inherit identical sets of genes, and they have a strong tendency to grow to the same height. Fraternal twins develop from separate eggs and are less likely to grow to the same height.

Delve Deeper

1. What are the possible outcomes to a protein if the gene that codes for it has a point mutation?
   It is possible that nothing would happen to the protein if the point mutation does not alter the sequence of amino acids. If the mutation does cause a different amino acid to be placed in the final protein, there is a good chance that the protein will not fold in the same way. This could either alter the function of the protein or make it completely nonfunctional.

2. Explain how a gene in one cell can influence the development of a different cell through the action of a hormone.
   The gene can code for a protein that will travel out of the cell and act as a signaling agent for the second cell. This protein can act as a signal for certain genes to turn on or off in the second cell, so that it will develop in a specific way dependent on its position.

3. What are two ways sexual reproduction can lead to offspring that are genetically different from their parents?
   Most organisms have many different chromosomes, each of which splits into maternal and paternal copies, and only one copy of each chromosome is transmitted to each offspring. Random combinations of maternal and paternal copies mix and match the alleles passed to offspring.

4. Why aren't the mutations that occur in skin cells or in other organs, such as the heart or brain, heritable?
   Because the cells dividing in the skin, heart, and brain are not involved in sexual reproduction. All these cells die when the individual dies. A mutation that occurs in these cells can't be passed down to offspring. Only eggs and sperm transmit information stored in DNA to the next generation, so only mutations that occur in these germ cells are heritable.

5. Why don't all phenotypic traits occur as discrete, alternative states like Mendel's peas?
   Phenotypic traits are rarely determined by single Mendelian loci. Besides the fact that variation in some traits can be attributed to the cumulative action of many genes, the environment can also influence phenotypic traits. Variation in the environment can lead to variation in the phenotypes that arise from a single genotype, as does the complex interactions between many different genes and the environment.

Test Yourself:

1. b; 2. c; 3. d; 4. e; 5. c; 6. b; 7. a; 8. e

Chapter 6
THE WAYS OF CHANGE: DRIFT AND SELECTION

Check Your Understanding

1. What is an organism's phenotype?
   a. The interaction of an organism's genes with the environment to produce characteristics such as how the amount of light a plant is exposed to influences its height.
   b. Characteristics of an organism that can be classified into discrete categories, such as gender or eye color.
   c. Any aspect of an organism that can be measured, such as how it looks, how it behaves, how it's structured.
   d. The genetic makeup of an individual.
   e. Both b and d.

2. How many alleles does a genetic locus have?
   a. One.
   b. Two.
   c. More than two.
   d. It depends on the locus.

3. Why is the variation of phenotypic traits often continuous, distributed around a mean in a bell-shaped curve?
   a. Because phenotypic traits are a result of dominance.
   b. Because phenotypic traits are not related to genotypes.
   c. Because phenotypic traits are only influenced by the environment.
   d. Because phenotypic traits are often polygenic.

Learning Objectives for Chapter 6:

- ➤ Explain Hardy-Weinberg equilibrium.
- ➤ Analyze the effects of genetic drift on large and small populations.
- ➤ Explain how drift and selection can interact during founder events to cause newly formed populations to differ substantially from their source population.
- ➤ Compare and contrast the kinds of knowledge that can be gained from laboratory versus "natural" studies of the evolution of E. coli.
- ➤ Demonstrate how deleterious mutations can persist within a population.
- ➤ Describe the differences between directional selection and stabilizing selection.
- ➤ Illustrate how selection and heritability affect the speed of evolution.

The Tangled Bank Study Guide

Identify Key Terms

Connect the following terms with their definitions on the right:

balancing selection	A type of genetic drift describing the loss of allelic variation that accompanies founding of a new population from a very small number of individuals (a small sample of a much larger source population). This can cause the new population to differ considerably from the source population.
directional selection	Selection that decreases the frequency of alleles within a population.
disruptive selection	A form of fitness that favors individuals with a trait near the mean of the population.
fitness	An allele that represents the only alternative for a particular genetic locus within a population; all members of that population are homozygous for that allele at that locus.
fixed allele	Evolution arising from random changes in the genetic composition of a population from one generation to the next.
founder effect	A form of fitness that favors individuals at one end of the distribution of a trait.
genetic bottleneck	Selection that increases the frequency of alleles within a population.
genetic drift	Event where the number of individuals in a population is reduced drastically. Even if this dip in numbers is temporary, it can have lasting effects on the genetic variation of a population.
genetic locus (plural: loci)	A mathematical principle that shows that in the absence of drift, selection, migration, and mutation, allele frequencies at a genetic locus will not change from one generation to the next.
Hardy-Weinberg equilibrium	A form of fitness that favors individuals at either end of the distribution.
negative selection	The specific location of a gene or piece of DNA sequence on a chromosome. When mutations modify the sequence at a locus, they generate new alleles—variants of a particular gene or DNA region. Alleles are mutually exclusive alternative states for a genetic locus.
positive selection	Selection that favors more than one allele. It acts to maintain genetic diversity in a population by keeping alleles at frequencies higher than would be expected by chance or mutation alone.
stabilizing selection	The success of an organism at surviving and reproducing and thus contributing offspring to future generations.

Link Concepts

1. Fill in the following concept map with key terms from the chapter showing the relationships among the genes, gene products, and outcomes in Hopi Hoekstra's research on oldfield mice.

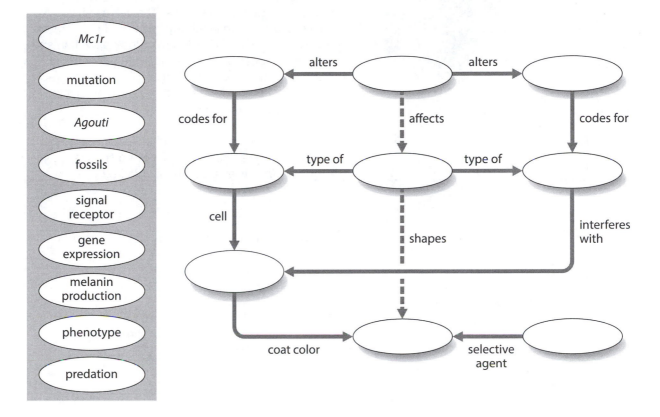

2. Develop a concept map that shows the relationships among resistance alleles in mosquitoes. Think about allele, cost, fitness, natural selection, resistance, and variation.

The Tangled Bank Study Guide

Interpret the Data

A

Selection on beak size

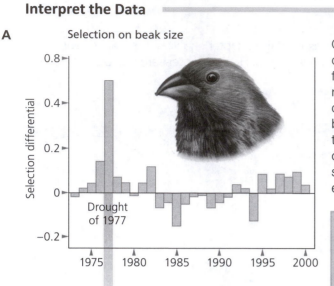

B

Evolution of beak size

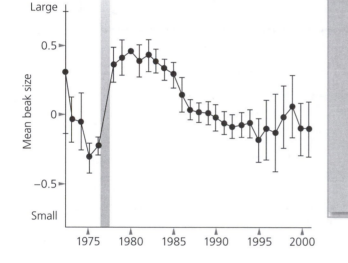

Over the past four decades, the strength and direction of natural selection on the beaks of finches in the Galápagos have fluctuated in response to available foods. The selection differential shown in the top graph is the difference between the mean beak size of the population and the mean beak size of the individuals producing offspring in the next generation. The bottom graph shows mean beak size (and the variation around each mean) for each year.

How many years did the selection differential favor large-billed finches (positive selection differential)?

Was variation in beak size similar in all years?

What does the shaded area in the lower graph indicate?

Games and Exercises

Asparagus Pee

Does your urine smell after you've eaten asparagus? A survey through 23andMe.com found that out of a sample of 4,737 individuals of European ancestry, 3,002 said they could smell asparagus in their urine and 1,735 said they could not. If the A allele for odor detection is dominant over the G allele for lack of odor detection, and 1,027 are heterozygotes, can you determine the allele frequencies for this locus, assuming random mating? What about the genotype frequencies? How do you know if the population is in Hardy-Weinberg equilibrium?

You must calculate both the allele and genotype frequencies using the data because you don't know whether this population meets Hardy-Weinberg assumptions. First, you have to figure out the frequency of the alleles. The frequency of the A allele will equal the total number of A alleles in the population relative to all the alleles in the population. So you have to calculate the number of alleles present in both the homozygotes and the heterozygotes. Because 1,027 individuals are heterozygous for the odor detection allele, 1,975 individuals must be homozygous (3,002 total that can smell the urine—1,027 heterozygotes). Since homozygous individuals carry two copies of the A allele, and heterozygous individuals carry one copy, the frequency of the A allele is 2 times the number of homozygotes plus 1 times the number of heterozygotes all divided by the total number of alleles in the population for the locus (the total number of individuals times 2 because each locus has two alleles):

$$f(A) = [(1975 \times 2) + 1027] / (4737 \times 2) =$$

The frequency of the G allele will equal the total number of G alleles in the population (homozygotes and heterozygotes) divided by the total number of alleles in the population for the locus:

$$f(G) = [(1735 \times 2) + 1027] / (4737 \times 2) =$$
$f(G)$ also equals $1 - f(A)$.

Second, to determine genotype frequencies, divide each genotype by the population size:

$$f(AA) = 1975/4737 =$$
$$f(AG) = 1027/4737 =$$
$$f(GG) = 1735/4737 =.$$

But if the Hardy-Weinberg assumptions are met, and no evolutionary mechanisms are operating on this locus, then the theorem predicts that the genotype frequencies should equal p^2, $2pq$, and q^2:

$$p^2 = 0.53^2 =$$
$$2pq = 2 \times 0.53 \times 0.47 =$$
$$q^2 = 0.47^2 =$$

The observed genotype frequencies do not equal those predicted by the Hardy-Weinberg theorem, but can you conclude that no evolutionary mechanisms operate on this locus in this population? Chi square is a statistical test used by population geneticists to determine whether the difference between observed and expected frequencies is significant—i.e., the likelihood that the difference may be due to chance. (Note: Chi square is calculated as the sum of each observed genotype frequency minus its expected frequency, divided by its expected frequency. The sum is compared to a distribution to determine significance.) In this case, the frequencies are statistically significant from expected.

Alan R. Lemmon developed a simulation of the Hardy-Weinberg theorem that shows what happens to genotype frequencies when assumptions are violated. The simulation also allows you to see the outcomes in different formats.

http://www.evotutor.org/EvoGen/EG1A.html

You can also see simulations of the effects of directional, disruptive, and stabilizing selection.

http://www.evotutor.org/Selection/Sl5A.html

Drifting Along

Kent Holsinger, professor of ecology and evolutionary biology at the University of Connecticut, put together a variety of simple simulations to help students understand principles of population genetics. Here are links to three of the simulations:

- Genetic drift—This simulation allows you to select different allele frequencies to start, different population sizes, and different numbers of generations for simulations.

  http://darwin.eeb.uconn.edu/simulations/drift.html

- Natural selection—This simulation allows you to select different fitness levels for each of three genotypes and view the outcome of natural selection after 100 generations.

  http://darwin.eeb.uconn.edu/simulations/selection.html

- Natural selection and genetic drift—This simple simulation allows you to examine the effects of both natural selection and drift by selecting from different starting allele frequencies, different population sizes, and different numbers of generations given a specific fitness, representing selection for an initially rare allele.

  http://darwin.eeb.uconn.edu/simulations/selection-drift.html

If you're interested in seeing more, go to:
http://darwin.eeb.uconn.edu/simulations/simulations.html

Go Online

Natural Selection—Crash Course Biology #14

In his "crash course" YouTube video, Hank Green discusses natural selection, adaptation, and fitness.

http://www.youtube.com/watch?v=aTftyFboC_M

Genetic Drift: Random Evolutionary Change

Paul Andersen shows how genetic drift can be a mechanism for evolutionary change in this YouTube video. He also discusses bottlenecks in northern elephant seals and the high incidence of total colorblindness that resulted from a typhoon that hit the small island of Pingelap.

http://www.youtube.com/watch?v=mjQ_yN5znyk

Lactase and Me

From genomicseducation, this YouTube video is an animation (complete with sound effects) that explains the genetics behind lactose intolerance.

http://www.youtube.com/watch?v=U4w-0qkYnjg

Galápagos: The Finches

Open University introduces the Galápagos finches in this YouTube video, including one species that uses tools to access food.

http://www.youtube.com/watch?v=l25MBq8T77w

Multiple Instances of Ancient Balancing Selection Shared between Humans and Chimpanzees

Ellen M. Leffler and her colleagues examined balancing selection in humans and chimpanzees by conducting genome-wide scans for sequence variations that differed by a single nucleotide between humans and chimpanzees (SNPs).

http://www.sciencemag.org/content/339/6127/1578.abstract?sid=9c531c35-0af9-4a0c-9e52-2cfda69e9322

Evolution in Our Own Time

In *Evolution via Roadkill*, *Science* Now reports on recent research that shows that cliff swallow wings are shorter than they were historically, largely because of the selective action of cars on swallow fitness.

http://news.sciencemag.org/sciencenow/2013/03/evolution-via-roadkill.html?ref=em

See the original report here:
http://www.cell.com/current-biology/retrieve/pii/S0960982213001942

Fluttering from the Ashes?

This News & Analysis from Science discusses how scientists are trying to identify genes that distinctly identify some extinct species, with the idea of bringing them back to life.

http://www.sciencemag.org/content/340/6128/19.summary

Common Misconceptions

Dominance Doesn't Necessarily Mean Dominance

Alleles are considered dominant when only a single copy is enough to produce a trait, but that doesn't mean that the dominant allele of a trait will always have the highest frequency in a population and the recessive allele will always have the lowest frequency. For example, Huntington's disease is a rare disease caused by an uncommon dominant allele, and blood type O is the most common blood type in humans but results from recessive alleles. Mendel's simple model crossing two heterozygous (Aa) parents yields a 3:1 ratio of A to a alleles in the offspring, so more of the dominant allele. But that cross is based on heterozygous individuals. If most individuals in a population are homozygous for the recessive allele (aa), then most offspring will be aa. Ultimately, allele frequencies are determined by natural selection and genetic drift.

Allele and Gene Cannot Be Used Interchangeably

An allele is one of any number of alternative forms of a gene or genetic locus, whereas a gene can be defined as segments of DNA whose nucleotide sequences code for proteins, or RNA, or regulate the expression of other genes. Although a gene is a complex concept (and the more scientists learn about DNA and gene expression the more difficult defining the concept becomes), an allele is a very specific difference in DNA coding that scientists are able to identify relative to other alleles.

Genetic Drift Occurs in Large Populations as Well as Small

Genetic drift is not just a phenomenon affecting small populations; all populations experience random changes in allele frequency from one generation to the next. The difference is that large populations simply have more individuals and likely more copies of any particular allele, so all the alleles are more likely to be represented in the next generation—not necessarily, of course.

Natural Selection and Genetic Drift Are Both Mechanisms for Evolution

Scientists debate over the role of natural selection and genetic drift in the evolution of organisms, but they agree that both are mechanisms for evolutionary change. Genetic drift is a nonselective process—it is random. But like natural selection, this random process does influence the frequency of alleles in the next generation.

Contemplate

How is the underlying genetic architecture of oldfield mouse coat color like the underlying genetic architecture of human height?	Why can't scientists conduct experiments like Richard Lenski's with organisms such as birds or mammals?

Delve Deeper

1. Why did the genetic variation of northern elephant seal populations remain low for thousands of generations after the bottleneck event? How could genetic drift have played a role in slowing the recovery of genetic diversity?

2. What would happen to the frequency of heterozygous carriers of sickle-cell anemia (with an AS genotype) if mosquitoes were completely wiped out in a large region? Explain.

3. If a mutation that produces a new allele that is deleterious arises in a population, what will most likely happen to the frequency of that allele?

4. What are the differences and similarities between directional and stabilizing selection?

5. What things did Peter and Rosemary Grant's team need to measure or record in order to demonstrate the effect of natural selection on the beak size of finches in the Galápagos?

6. In order for evolution by natural selection to occur, why is it important for the coat color of oldfield mice to be variable and at least partly heritable? What would happen if the variation or heritability were reduced?

7. Why has the evolution of resistance to insecticides in mosquitoes been so rapid?

Go the Distance: Examine the Primary Literature

Sash Vignieri, Joanna G. Larson, and Hopi E. Hoekstra test the effects of crypsis on survival of oldfield mice in their recent report in the journal Evolution.

- What predictions did these scientists make about crypsis?

- How did they test their predictions?

- Why do they argue that stabilizing selection maintains color matching within a particular habitat?

Vignieri, S. N., J. G. Larson, and H. E. Hoekstra. 2010. The Selective Advantage of Crypsis in Mice. *Evolution* 64 (7): 2153–58. doi:10.1111/ j.1558-5646.2010.00976.x. http://www.oeb.harvard.edu/faculty/hoekstra/PDFs/Vignieri2010Evol.pdf.

Test Yourself

1. If two individuals mate, one of which is heterozygous for a dominant allele at a locus and the other is homozygous for a recessive allele at the same locus, what will be the outcome?
   a. The offspring will be either heterozygous or homozygous for the recessive allele.
   b. The offspring will be either homozygous for the dominant allele, heterozygous, or homozygous for the recessive allele.
   c. The offspring will not evolve because they will carry the same alleles as the parents.
   d. The recessive allele will eventually become the dominant allele in the population.
   e. None of the above.

2. The Hardy-Weinberg theorem is an important mathematical proof because:
   a. It demonstrates that dominant alleles are more common than recessive alleles.
   b. It demonstrates that allele frequencies of a population will not change from one generation to the next in the absence of outside forces.
   c. It demonstrates that a locus can only have one of two alleles.
   d. It demonstrates that heterozygotes are always better.
   e. It demonstrates evolution.

3. Which population would be most likely to have allele frequencies in Hardy-Weinberg equilibrium?
   a. A population in a rapidly changing environment.
   b. A population where immigration is common.
   c. A population that is currently not evolving.
   d. A population that cycles between a very large and very small number of individuals.

4. Which of the following statements is/are not true about genetic drift?
   a. Genetic drift can cause the loss of an allele in a species.
   b. Genetic drift happens faster in large populations than small ones.
   c. Genetic drift causes evolution of a population.
   d. Both a and b.

   e. Both b and c.

5. A genetic bottleneck in a population will often result in what?
   a. Loss of alleles.
   b. Loss of genetic diversity.
   c. An increase in genetic drift.
   d. All of the above.
   e. None of the above.

6. When will an allele with a higher fitness than any other allele at the same locus become more common?
   a. Always.
   b. Never.
   c. It depends only on the strength of selection.
   d. It depends on both the strength of selection and the size of the population.

7. Which of the following statements about balancing selection is true?
   a. It is only possible in sexually reproducing species.
   b. It is responsible for maintaining the S allele for sickle-cell anemia within humans.
   c. It is not possible when heterozygotes have a higher fitness.
   d. It is the reason why sickle-cell anemia is selected against in areas with high levels of malaria.
   e. None. All are false statements.

8. Why did the third generation of Hoekstra's mice (2nd generation hybrids) develop coat colors that ranged from nearly all white to mostly brown?
   a. Because coat color is a result of the environment, and individuals in beach environments need to be light while individuals in inland environments need to be brown.
   b. So that some offspring will always match their environment because a range of coat colors is necessary for individuals to survive in different environments.
   c. Because coat color is a polygenic trait, and each of the mice in the third generation had a

different combination of genetic regions influencing coloration.

d. Because as dominant alleles for coat color become rarer in the offspring as a result of

hybridization, the recessive alleles influence coat color more and more.

e. None of the above.

Answers for Chapter 6
Check Your Understanding

1. What is an organism's phenotype?

   a. The interaction of an organism's genes with the environment to produce characteristics such as how the amount of light a plant is exposed to influences its height.

   Incorrect. An organism's genes do interact with the environment (both external and internal environments) during development to produce phenotypic characteristics, but the phenotype is the characteristic itself. The phenotype is any aspect of an organism that can be measured such as morphology, physiology, and behavior (see Chapter 5).

   b. Characteristics of an organism that can be classified into discrete categories, such as gender or eye color.

   Incorrect. Some phenotypic characteristics can be classified into discrete categories, such as gender (female or male) or eye color (blue, green, brown), but many other characteristics cannot, such as height, length of the femur, antler size (these traits vary in a range around a mean). The phenotype is any aspect of an organism that can be measured such as morphology, physiology, and behavior (see Chapter 5).

   c. Any aspect of an organism that can be measured, such as how it looks, how it behaves, how it's structured.

   Correct. The phenotype is linked to the genotype—genes interact with other genes and with the environment during the development of the phenotype. Some of those aspects can be classified into discrete categories, such as gender (female or male) or eye color (blue, green, brown), but many other characteristics cannot, such as height, length of the femur, antler size (these traits vary in a range around a mean). See Chapter 5 for a discussion of the complexities of the links between phenotypes and genotypes.

   d. The genetic makeup of an individual.

   Incorrect. An organism's phenotype is any measurable aspect, such as morphology, physiology, or behavior. The genetic makeup of an individual is that individual's genotype. See Chapter 5 for a discussion of the complexities of the links between genotypes and phenotypes.

   e. Both b and d.

   Incorrect. Some phenotypic characteristics can be classified into discrete categories, such as gender (female or male) or eye color (blue, green, brown), but many other characteristics cannot, such as height, length of the femur, antler size (these traits vary in a range around a mean). The phenotype is any aspect of an organism that can be measured such as morphology, physiology, and behavior, and the genetic makeup of an individual is that individual's genotype (see Chapter 5).

2. How many alleles does a genetic locus have?

   a. One.

   Incorrect. Fixed alleles, by definition, are the only allele at a particular genetic locus within a population; all members of that population are homozygous for that allele at that locus. Fixed loci are not necessarily typical, however; there are no hard and fast rules about the number of alleles at any particular genetic locus. Mutations can create new alleles at any genetic locus, and the number of alleles and their frequency within a population can change as a result of natural selection and genetic drift.

   b. Two.

   Incorrect. Simple models of loci often represent alleles in upper- and lowercase letters, such as AA, Aa, and aa, implying that a locus only has two alleles. The number of alleles at any locus is not limited to two, however, and more accurate models assign allele names that reflect the option for greater diversity. There are no hard and fast rules about the number of alleles at any particular genetic locus, however. Mutations can create new alleles at any genetic locus, and the number of alleles and their frequency within a population persist or disappear as a result of natural selection and genetic drift.

   c. More than two.

   Incorrect. There are no hard and fast rules about the number of alleles at any particular genetic locus. Mutations can create new alleles at any genetic locus, and the number of alleles and their frequency within a population persist or disappear as a result of natural selection and genetic drift. Although the simple models of loci imply that a locus only has two alleles because they use upper- and lowercase letters, such as AA, Aa, and aa, that doesn't necessarily mean that genetic locus has only two alleles. More accurate models assign allele names that reflect the option for greater diversity of alleles at any particular locus—whether that be one, two, or more than two.

d. It depends on the locus.

Correct. There are no hard and fast rules about the number of alleles at any particular genetic locus. Mutations can create new alleles at any genetic locus, and the number of alleles and their frequency within a population can change as a result of natural selection and genetic drift. A more appropriate way of representing alleles incorporates subscripts (A_1, A_2, A_3, A_4) or superscripts (Ester[1], Ester[4]), reflecting the diverse number of alleles that may or may not be known (yet) for any particular locus.

3. Why is the variation of phenotypic traits often continuous, distributed around a mean in a bell-shaped curve?

a. Because phenotypic traits are a result of dominance.

Incorrect. Dominance can affect phenotypic traits, but dominance refers to alleles at a single genetic locus. Although Mendel was able to predict some phenotypes, most phenotypic traits cannot be reduced to the simple patterns of inheritance Mendel studied. Most phenotypic traits involve more than one gene (i.e., they are polygenic), and the different alleles from these different genes contribute to the trait differently. The result—a continuous distribution (like height, from 5'5" to 5'5 ¼" to 5'5 ½" and all the possible measures in between). Most individuals will have measurements near the mean, with fewer and fewer in the "tails," so the curve appears bell shaped.

b. Because phenotypic traits are not related to genotypes.

Incorrect. Although the relationship between genotype and phenotype is not necessarily straightforward, the genotype and phenotype are in fact intimately related. Most phenotypic traits involve more than one gene (i.e., they are polygenic), and the different alleles from these different genes contribute to the trait differently. The result—a continuous distribution (like height, from 5'5" to 5'5 ¼" to 5'5 ½" and all the possible measures in between). Most individuals will have measurements near the mean, with fewer and fewer in the "tails," so the curve appears bell shaped.

c. Because phenotypic traits are only influenced by the environment.

Incorrect. The environment is indeed an important component of the expression of phenotypic traits, but a distinct genetic architecture underlies that influence. Indeed, most phenotypic traits involve more than one gene (i.e., they are polygenic), and the different alleles from these different genes contribute to the trait differently. The result—a continuous distribution (like height, from 5'5" to 5'5 ¼" to 5'5 ½" and all the possible measures in between). Most individuals will have measurements near the mean, with fewer and fewer in the "tails," so the curve appears bell shaped.

d. Because phenotypic traits are often polygenic.

Correct. Most phenotypic traits involve more than one gene (i.e., they are polygenic), and the different alleles from these different genes contribute to the trait differently. The result—a continuous distribution (like height, from 5'5" to 5'5 ¼" to 5'5 ½" and all the possible measures in between). Most individuals will have measurements near the mean, with fewer and fewer in the "tails," so the curve appears bell shaped.

Identify Key Terms

balancing selection	Selection that favors more than one allele. It acts to maintain genetic diversity in a population by keeping alleles at frequencies higher than would be expected by chance or mutation alone.
directional selection	A form of fitness that favors individuals at one end of the distribution of a trait.
disruptive selection	A form of fitness that favors individuals at either end of the distribution.
fitness	The success of an organism at surviving and reproducing and thus contributing offspring to future generations.
fixed allele	An allele that represents the only alternative for a particular genetic locus within a population; all members of that population are homozygous for that allele at that locus.
founder effect	A type of genetic drift describing the loss of allelic variation that accompanies founding of a new population from a very small number of individuals (a small sample of a much larger source population). This can cause the new population to differ considerably from the source population.
genetic bottleneck	Event where the number of individuals in a population is reduced drastically. Even if this dip in numbers is temporary, it can have lasting effects on the genetic variation of a population.
genetic drift	Evolution arising from random changes in the genetic composition of a population from one generation to the next.
genetic locus (plural: loci)	The specific location of a gene or piece of DNA sequence on a chromosome. When mutations modify the sequence at a locus, they generate new alleles—variants of a particular gene or DNA region. Alleles are mutually exclusive alternative states.
Hardy-Weinberg equilibrium	A mathematical principle that shows that in the absence of drift, selection, migration, and mutation, allele frequencies at a genetic locus will not change from one generation to the next.
negative selection	Selection that decreases the frequency of alleles within a population.
positive selection	Selection that increases the frequency of alleles within a population.
stabilizing selection	A form of fitness that favors individuals with a trait near the mean of the population.

Link Concepts

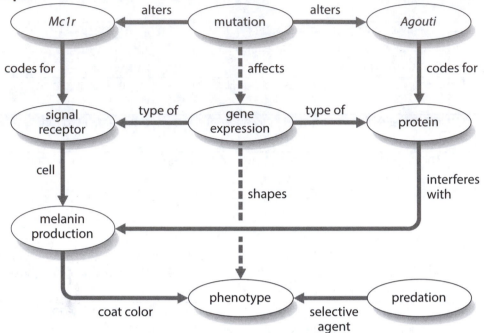

Interpret the Data

- How many years did the selection differential favor large-billed finches (positive selection differential)?

  16 years.

- Was variation in beak size similar in all years?

  No. In some years, the bars surrounding the mean in the lower graph are short (little variation) and in others the bars are long (great variation).

- What does the shaded area in the lower graph indicate?

  Rapid evolutionary response to a particularly strong episode of directional selection. The area highlights a large and significant change in mean beak size for the finch population in a very short time frame.

Games and Exercises

$f(A) = [(1975 \times 2) + 1027] / (4737 \times 2) = 0.53$.
$f(G) = [(1735 \times 2) + 1027] / (4737 \times 2) = 0.47$.
$f(AA) = 1975/4737 = 0.42$
$f(AG) = 1027/4737 = 0.22$
$f(GG) = 1735/4737 = 0.37$.
$p^2 = 0.53^2 = 0.28$
$2pq = 2 \times 0.53 \times 0.47 = 0.50$
$q^2 = 0.47^2 = 0.23$.

Delve Deeper

1. Why did the genetic variation of northern elephant seal populations remain low for thousands of generations after the bottleneck event? How could genetic drift have played a role in slowing the recovery of genetic diversity?

Many alleles were lost in the bottleneck because of the death of the individuals that carried those alleles—they literally could not reproduce and pass those alleles to their offspring. The small number of survivors had only a subset of the original genetic diversity of the larger population. Because the recovering populations were very small, genetic drift had a powerful effect on variation. Not only were alleles present in the population lost, but a new mutation that did arise was also more likely to be lost. Ultimately, as the populations began to recover, the population size increased faster than the rate at which new mutations arose in the population, and genetic diversity remained low.

2. What would happen to the frequency of heterozygous carriers of sickle-cell anemia (with an AS genotype) if mosquitoes were completely wiped out in a large region? Explain.

The frequency of AS individuals would decrease without balancing selection. Without mosquitoes there would be no transmission of malaria and so the S allele would lose its advantage of protecting against death from malaria. It would not likely disappear right away because as the S allele became rarer and rarer, the probability of it occurring in a homozygous genotype (i.e., paired with another S allele) would be low. Once in this rare state, drift alone would determine whether it persisted in the population. Selection could only act on the S allele when in the homozygous SS genotype, which causes sickle-cell anemia in the phenotype.

3. If a mutation that produces a new allele that is deleterious arises in a population, what will most likely happen to the frequency of that allele?

Not all deleterious mutations disappear from populations; the outcome depends on the allele's effect on the phenotypes. If the mutant allele is recessive, it may be rare enough that it is almost never expressed in a homozygous state. It can even remain at a low frequency within the population for a very long time. Of course, drift may also determine whether the allele persists in the population.

4. What are the differences and similarities between directional and stabilizing selection?

Both require the same three conditions for evolution in response to natural selection: variation, heritability, and some phenotypes having a higher survival or reproduction than others. They differ because directional selection favors increases, or decreases, in the size or dimensions of a trait, while stabilizing selection favors an intermediate value for the trait. Directional selection will always change the average value of a trait in a population (provided there is variation), while stabilizing selection will tend to keep it the same.

5. What things did Peter and Rosemary Grant's team need to measure or record in order to demonstrate the effect of natural selection on the beak size of finches in the Galápagos?

They needed to measure the beak size of many individuals in multiple generations to determine how this trait is inherited and how it changes from one generation to the next. They also had to measure the size of the seeds that the different birds were eating to understand the selection pressures on beak size. By marking individuals, they could determine exactly who survived and reproduced.

6. In order for evolution by natural selection to occur, why is it important for the coat color of oldfield mice to be variable and at least partly heritable? What would happen if the variation or heritability were reduced?

If all individuals were identical, no particular individual should survive or reproduce any better than any other. But if individuals vary in a trait such as coat color, some individuals may be able to survive or reproduce better than others. If the coat color wasn't at least partly heritable, the traits of the survivors could not be passed on to their offspring. If the variation or the heritability were reduced, the effect of natural selection would be slowed and the evolution of coat color in the mice would not happen as quickly.

7. Why has the evolution of resistance to insecticides in mosquitoes been so rapid? Insecticides, and other pesticides, can impose extremely strong selection. Pest populations often are highly genetically variable. They don't necessarily have mutation rates any higher than other organisms; the variation comes from tremendous reproductive rates. As a result, mutations that are heritable can have huge effects on the next generation and resistance can evolve within the population rather rapidly.

Test Yourself:
1. a; 2. b; 3. c; 4. b; 5. d; 6. d; 7. b; 8. c

Chapter 7
MOLECULAR EVOLUTION: THE HISTORY IN OUR GENES

Check Your Understanding

1. Which of the following statements about genetic drift is TRUE?
   a. Genetic drift occurs when populations move to new locations.
   b. Because species need specific adaptations to survive, genetic drift cannot be an important factor in evolution.
   c. Genetic drift is a random process that contributes to evolutionary change within populations.
   d. Genetic drift is a form of natural selection.
   e. Genetic drift refers to the random mutations that affect small populations.

2. Which of the follow are NOT noncoding regions of DNA?
   a. Pseudogenes.
   b. Proteins.
   c. MicroRNAs.
   d. Mobile genetic elements.

3. Why is heritable variation among individuals an important factor for natural selection?
   a. Because variation has to be inherited from parent to offspring for a species to survive.
   b. Because variation has to be heritable for individuals to pass down their beneficial mutations.
   c. Because when individuals respond to the environment, they can pass that success on to offspring.
   d. Because natural selection cannot act when all individuals are absolutely identical.
   e. Both b and d.

Learning Objectives for Chapter 7:

➤ Explain how and why gene trees can be different from species trees.
➤ Describe four examples of how DNA has been used to examine phylogenetic relationships.
➤ Compare and contrast neutral evolution and natural selection.
➤ Explain how molecular clocks can be used to deduce time.
➤ Compare the effects of positive and purifying selection on synonymous and replacement substitutions within a pseudogene.

Identify Key Terms

Connect the following terms with their definitions on the right:

coalescence	Noncoding stretches of DNA containing strings of short (1–6 base pairs), repeated segments. The number of repetitive segments can be highly polymorphic, so these segments serve as valuable genetic characters for comparing populations and for assigning relatedness among individuals (DNA fingerprinting).
conserved region	A theory that describes the pattern of nucleotide sequence evolution under the forces of mutation and random genetic drift in the absence of selection. Under these conditions, the rate at which one nucleotide replaces another in a population (the substitution rate) should equal the rate of mutation at that site (mutation rate) regardless of the size of the population.
gene tree	Selection that removes deleterious alleles, such as mutations to non-synonymous sites that disrupt essential functions.
microsatellite	The process by which, looking back through time, the genealogy of any pair of homologous alleles merges in a common ancestor.
molecular clock	A functional segment of DNA that is almost identical among every species in which scientists have looked.
neutral theory of molecular evolution	Mutations that alter the amino acid sequence of a protein. These can affect the phenotype and are therefore more subject to selection.
positive selection	Mutations that do not alter the amino acid sequence of a protein. Because these mutations do not affect the protein an organism produces, they are less prone to selection and often free from selection completely.
purifying selection	The branched genealogical lineage of homologous alleles, which traces their ancestry back to an ancestral allele.
replacement substitution	Selection that increases the frequency of alleles within a population, such as mutations to non-synonymous sites that enhance essential functions.
species tree	A method used to determine time based on base pair substitutions. Rates of change can be used to deduce the divergence time between two lineages in a phylogeny, for example. Other markers of time, such as fossils with known ages, can be used to calibrate this methodology.
synonymous substitution	The branched lineage of species, which traces the evolutionary history back to a common ancestor.

Link Concepts

1. Fill in the following concept map with the key terms from the chapter:

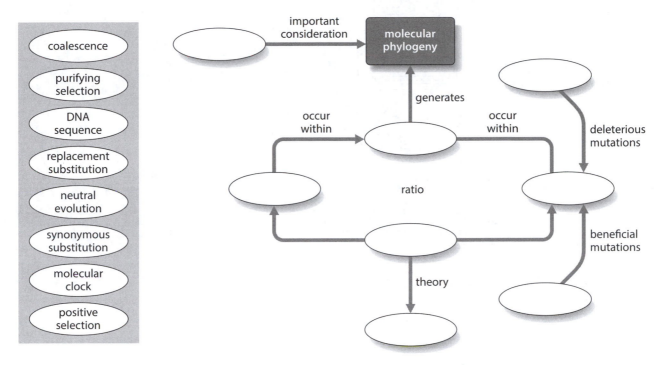

2. Develop your own concept map that explains the evolution of HIV, using the concepts HIV, SIV, mutation, chimpanzee, human, amino acid, and protein.

Interpret the Data

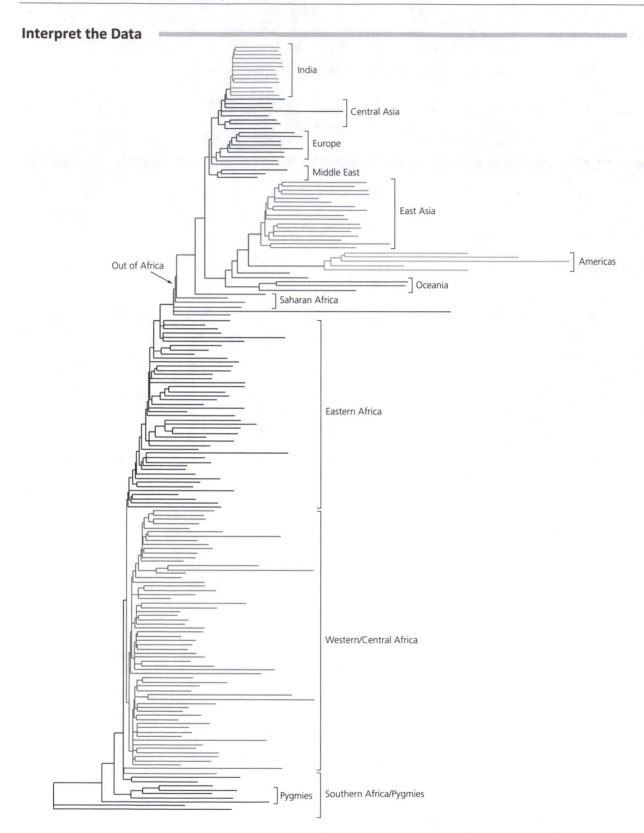

Sarah Tishkoff and colleagues developed this phylogeny of humans based on mitochondrial DNA (Figure 7.7).

What group shares the most recent common ancestor with the group of humans that colonized the Americas?

According to Tishkoff et al. (2009), the longer branch lengths for groups of humans in the Americas, Oceania, and Pygmy, and some of the hunter-gatherers, indicate high levels of genetic drift. Why might those groups have experienced higher levels of genetic drift than groups that remained in Africa?

Games and Exercises

More Fun with Cladograms

In Chapter 4, using some simple characters, you were able to develop a cladogram showing the relationships of cartoon beetles. Although it may have seemed a bit silly to do this with cartoons, scientists use similar methods to examine the relationships among genes, species, and even languages! In fact, Walter Fitch and Charles Langley were able to plot out the evolutionary changes in a gene that codes for a protein called cytochrome *c*. Cytochrome *c* is necessary for energy transfer in cells and is found in most, if not all, eukaryotes. Fitch and Langley worked out the mutations that had accumulated in 17 lineages, including humans and horses, and found that the mutations became fixed with clocklike regularity.

 With this exercise developed by Beth Kramer (and added to ENSI website in 2002, http://www.indiana.edu/~ensiweb/lessons/mol.bio.html), you can compare the relatedness of a variety of eukaryotes—including humans—based on cytochrome *c* amino acid sequences.

Start by comparing the amino acid sequences of cytochrome *c* between pairs of species. Use horses, donkeys, whales, chickens, penguins, snakes, moths, yeast, and wheat (see Table 1). You aren't comparing the DNA sequences themselves—codons are associated with the specific amino acids (see Chapter 5), and you are comparing amino acids.

Make sure you examine the entire molecule. Each protein sequence has 103–12 amino acids. The sequences start in the top half and continue onto a second line in the bottom half. Count the total number of differences between species pairs (the shaded amino acids are the same in all species) and record them in Table 1. Note also that some species have amino acids in locations that other species don't—those also count as differences.

Remember, you only have to fill out one-half of the table because the halves mirror each other.

	Horse	Donkey	Whale	Chicken	Penguin	Snake	Moth	Yeast	Wheat
Horse	0								
Donkey		0							
Whale			0						
Chicken				0					
Penguin					0				
Snake						0			
Moth							0		
Yeast								0	
Wheat									0

Amino Acid Symbols

Symbol	Amino Acid
A	Alanine
C	Cysteine
D	Aspartic acid
E	Glutamic acid
F	Phenylalanine
G	Glycine
H	Histidine
I	Isoleucine
K	Lysine
L	Leucine
M	Methionine
N	Asparagine
P	Proline
Q	Glutamine
R	Arginine
S	Serine
T	Threonine
V	Valine
W	Tryptophan
Y	Tyrosine

Sequence alignment (Amino Acid positions 1–59)

Amino Acid Number	Sequence (1 … 10 … 20 … 30 … 40 … 50 …)
Human, Chimpanzee	- - - - - - - - G D V E K G K K I F I M K C S Q C H T V E K G G K H K T G P N L H G L F G R K T G Q A P G V S Y T A A
Rhesus monkey	- - - - - - - - G D V E K G K K I F I M K C S Q C H T V E K G G K H K T G P N L H G L F G R K T G Q A P G V S Y T A A
Horse	- - - - - - - - G D V E K G K K I F V Q K C A Q C H T V E K G G K H K T G P N L H G L F G R K T G Q A P G F T Y T D A
Donkey	- - - - - - - - G D V E K G K K I F V Q K C A Q C H T V E K G G K H K T G P N L H G L F G R K T G Q A P G F S Y T D A
Common zebra	- - - - - - - - G D V E K G K K I F V Q K C A Q C H T V E K G G K H K T G P N L H G L F G R K T G Q A P G F S Y T D A
Pig, Cow, Sheep	- - - - - - - - G D V E K G K K I F V Q K C A Q C H T V E K G G K H K T G P N L H G L F G R K T G Q A P G F S Y T D A
Dog	- - - - - - - - G D V E K G K K I F V Q K C A Q C H T V E K G G K H K T G P N L H G L F G R K T G Q A P G F S Y T D A
Gray whale	- - - - - - - - G D V E K G K K I F V Q K C A Q C H T V E K G G K H K T G P N L H G L F G R K T G Q A P G F S Y T D A
Rabbit	- - - - - - - - G D V E K G K K I F V Q K C A Q C H T V E K G G K H K T G P N L H G L F G R K T G Q A V G F S Y T D A
Kangaroo	- - - - - - - - G D V E K G K K I F V Q K C A Q C H T V E K G G K H K T G P N L H G L F G R K T G Q A P G F T Y T D A
Chicken, Turkey	- - - - - - - - G D I E K G K K I F V Q K C S Q C H T V E K G G K H K T G P N L H G L I G R K T G Q A E G F S Y T D A
Penguin	- - - - - - - - G D I E K G K K I F V Q K C S Q C H T V E K G G K H K T G P N L H G L I G R K T G Q A E G F S Y T D A
Peking duck	- - - - - - - - G D V E K G K K I F V Q K C S Q C H T V E K G G K H K T G P N L H G L F G R K T G Q A E G F S Y T D A
Snapping turtle	- - - - - - - - G D V E K G K K I F V Q K C S Q C H T V E E G G K H K T G P N L N G L I G R K T G Q A E G F S Y T E A
Rattlesnake	- - - - - - - - G D V E K G K K I F S M K C S T C H T V E K G G K H K T G P N L H G L F G R K T G Q A P G F S Y S N A
Bullfrog	- - - - - - - - G D V E K G K K I F V Q K C A Q C H T C E K G G K H K V G P N L Y G L I G R K T G Q A A G F S Y T D A
Tuna	- - - - - - - - G D V A K G K K T F V Q K C A Q C H T V E N G G K H K V G P N L W G L F G R K T G Q A E G Y S Y T D A
Screwworm fly	- - - - - - G V P A G D V E K G K K I F V Q R C A Q C H T V E A G G K H K V G P N L H G L I G R K T G Q A A G F A Y T N A
Silkworm moth	- - - - - - G V P A G N A E N G K K I F V Q R C A Q C H T V E A G G K H K V G P N L H G F Y G R K T G Q A P G F S Y S N A
Tomato hoen "worm"	- - - - A S F S E A P P G N A D N G K K I F K T K C P Q C H T V D A G A G H K Q G P N L H G L F G R Q S G T T A G Y S Y S A A
Wheat	A S F S E A P P G N P K A G E K I F K T K C A Q C H T V D K G A G H K Q G P N L H G L F G R Q S G T T P G Y S Y S T A
Rice	A S F S E A P P G N P D A G A K I F K T K C A Q C H T V D A G A G H K Q G P N L H G L F G R Q S G T T A G Y S Y S T A
Baker's yeast (Fungus)	- - T E F K A G S A K K G A T L F K T R C L Q C H T V E K G G P H K V G P N L H G I F G R H S G Q A E G Y S Y T D A
Candida yeast (Fungus)	- - P A P F E Q G S A K K G A T L F K T R C A E C H T I E A G G P N K V G P N L H G I F S R H S G Q A Q G Y S Y T D A
Neurospora (Fungus)	- - - G F S A G D S K K G A N L F K T R C S E C H T E G G N L T Q K I G P A L H G L F G R K T G Q V D G Y A Y T D A

Sequence alignment (continued from above, Amino Acid positions 60–112)

Amino Acid Number	Sequence (60 … 70 … 80 … 90 … 100 … 110 …)
Human, Chimpanzee	N K N K G I T W G E D T L M E Y L E N P K K Y I P G T K M I F V G I K K K E E R A D L I A Y L K K A T N E
Rhesus monkey	N K N K G I T W G E D T L M E Y L E N P K K Y I P G T K M I F V G I K K K E E R A D L I A Y L K K A T N E
Horse	N K N K G I T W K E E T L M E Y L E N P K K Y I P G T K M I F A G I K K K T E R E D L I A Y L K K A T N E
Donkey	N K N K G I T W K E E T L M E Y L E N P K K Y I P G T K M I F A G I K K K T E R E D L I A Y L K K A T N E
Common zebra	N K N K G I T W K E E T L M E Y L E N P K K Y I P G T K M I F A G I K K K T E R E D L I A Y L K K A T N E
Pig, Cow, Sheep	N K N K G I T W G E E T L M E Y L E N P K K Y I P G T K M I F A G I K K K G E R E D L I A Y L K K A T N E
Dog	N K N K G I T W G E E T L M E Y L E N P K K Y I P G T K M I F A G I K K T G E R A D L I A Y L K K A T K E
Gray whale	N K N K G I T W G E E T L M E Y L E N P K K Y I P G T K M I F A G I K K K G E R E D L I A Y L K K A T N E
Rabbit	N K N K G I T W G E D T L M E Y L E N P K K Y I P G T K M I F A G I K K K D E R A D L I A Y L K K A T N E
Kangaroo	N K N K G I T W G E D T L M E Y L E N P K K Y I P G T K M I F A G I K K K G E R A D L I A Y L K K A T N E
Chicken, Turkey	N K N K G I T W G E D T L M E Y L E N P K K Y I P G T K M I F A G I K K K S E R V D L I A Y L K D A T S K
Penguin	N K N K G I T W G E D T L M E Y L E N P K K Y I P G T K M I F A G I K K K S E R A D L I A Y L K D A T S K
Peking duck	N K N K G I T W G E D T L M E Y L E N P K K Y I P G T K M I F A G I K K K A E R A D L I A Y L K D A T A K
Snapping turtle	N K N K G I T W G E E T L M E Y L E N P K K Y I P G T K M I F A G I K K K A E R A D L I A Y L K D A T S K
Rattlesnake	N K N K G I T W G E D T L M E Y L E N P K K Y I P G T K M V F T G L K K K T E R T D L I A Y L K E A T A K
Bullfrog	N K N K G I T W G E D T L M E Y L E N P K K Y I P G T K M V F A G L K K P N E R Q D L I A Y L K S A C S K
Tuna	N K S K G I V W N D D T L M E Y L E N P K K Y I P G T K M V F A G L K K A N E R Q D L V A Y L K S A T K -
Screwworm fly	N K A K G I T W Q D D T L F E Y L E N P K K Y I P G T K M V F A G L K K P N E R G D L I A Y L K S A T K -
Silkworm moth	N K A K G I T W Q D D T L F E Y L E N P K K Y I P G T K M V F A G L K K A N E R A D L I A Y L K E S T K -
Tomato hoen "worm"	N K A A V E W E E N T L Y D Y L L N P K K Y I P G T K M V F P G L K K P Q D R A D L I A Y L K Q A T K -
Wheat	N K N M A V I W E E N T L Y D Y L L N P K K Y I P G T K M V F P G L K K P Q E R A D L I S Y L K E A T S -
Rice	N K N M A V I W E E N T L Y D Y L L N P K K Y I P G T K M V F P G L K K P Q E R A D L I S Y L K E A T S -
Baker's yeast	N I K K N V O W D E N N M S E Y L T N P K K Y I P G T K M A F G G L K K E K D R N D L I T Y L K K A C E -
Candida yeast	N K R A G V E W A E P T M S D Y L L N P K K Y I P G T K M A F G G L K K D K D R N D L I T Y M L E A S A -
Neurospora (Fungus)	N K O K G I T W D E N T L F E Y L E N P K K Y I P G T K M A F G G L K K D K D R N D I I T F M K E A T A -

Use the data in Table 1 to make a cladogram. The most closely related species have the fewest differences in amino acid sequences. Examine the table and determine which species are most closely related. Use the tree below and place species according to their relationships. Start with the two most closely related. Add those species to the shortest lines in the tree to indicate they are the most closely related. Is there another species pair that is closely related? Add them to the next shortest lines for the same reason. Continue to add species based on their relative relatedness until all have been placed.

Why do more closely related organisms have more similar cytochrome *c*?

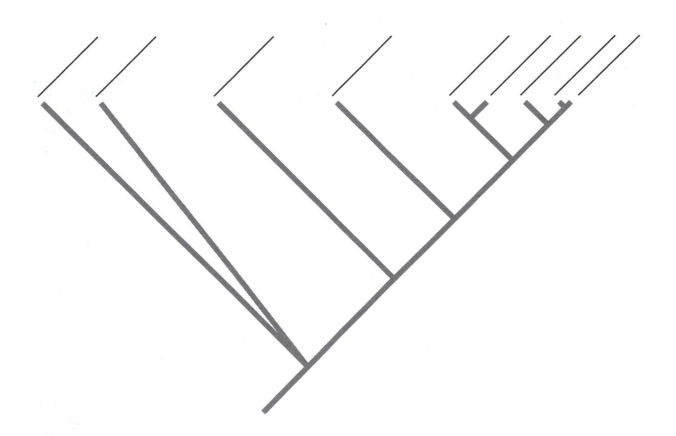

Now try adding the cytochrome c sequence difference data for humans.

	Horse	Donkey	Whale	Chicken	Penguin	Snake	Moth	Yeast	Wheat
Human									

How confident are scientists in these relationships? Why or why not?

What other tools do they use to determine the relationships among these species?

Go Online

 Bird Brains

WGBH and NOVA Science Now present the evidence for the evolution of the *FOXP2* gene, its effect on birdsong evolution and what that means for understanding human speech. The site includes a Q&A with Ofer Tchernichovski, a scientist studying vocal learning in birds at the City College of New York, a matching game for birds and their songs, and a great audio slide show that shows how experts are starting to learn to speak "walrus."

http://www.pbs.org/wgbh/nova/nature/bird-brains.html

 Jake Westhoff—Molecular Clocks

Jake Westhoff sings about molecular clocks, using them as a metaphor for human relationships in this YouTube video.

http://www.youtube.com/watch?v=PEUNTk75ESo

Pecking Away at Beak Evolution

Science Now discusses two important research reports from Arhat Abzhanov and colleagues and Ping Wu and colleagues that together illuminate the molecular basis of the species diversity of Darwin's finches on the Galápagos Islands.

http://news.sciencemag.org/sciencenow/2004/09/02-03.html?ref=hp

You can view the original research papers here:

Bmp4 and Morphological Variation of Beaks in Darwin's Finches
http://www.sciencemag.org/cgi/content/full/305/5689/1462

 Molecular Shaping of the Beak
http://www.sciencemag.org/cgi/content/full/305/5689/1465

Common Misconceptions

Mutations Can Be Deleterious, Beneficial, and Even Neutral.

Mutations are errors that arise when DNA is replicated, but just because they are errors doesn't always make them harmful. Most may be neutral, or synonymous substitutions. In fact, recent research on humans indicates that from 20 to 40 mutations occur in each gamete (http://www.nature.com.weblib.lib.umt.edu:8080/ng/journal/v43/n7/full/ng.862.html)! New mutations lead to variation among individuals (and they generate genetic variation within populations). And while mutations may occur randomly, their persistence in a population is a result of the mechanisms of evolution, such as natural selection and genetic drift.

Tips and Relatedness

Think of a phylogenetic tree as a mobile, with each group of branches spinning freely at each node. The tips aren't locked in to where they appear on the page. Relatedness comes from the shared common ancestors, i.e., the nodes.

Exactly What Do Straight Lines in Phylogenies Mean?

Phylogenies illustrate relationships among groups of organisms (e.g., genes, populations, or species). Most say very little about the amount of evolutionary change that has occurred along the length of a particular branch, however; branch length is often just a function of the order of the tree. The long straight branches don't mean a species hasn't changed. In fact, in some phylogenies, they may mean more evolutionary change within a lineage. As Chapter 7 explains, mutations are occurring all the time. Purifying selection tends to remove deleterious mutations from populations, but other mutations can exist simply because they are neutral (e.g., if they are synonymous substitutions). Neutral and positive mutations effect can accumulate in a population's genome, even though no lineages ever split.

Time Is Relative—When It's Looked at Correctly.

No matter how a phylogeny is drawn, reading time simply depends on the location of the common ancestor. For example,

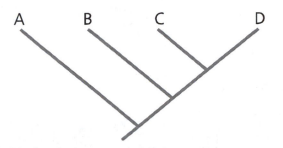

time in this style of tree is represented from bottom to top—the extant (living) species or groups are listed at the top, and each successive node down represents an earlier and earlier common ancestor.

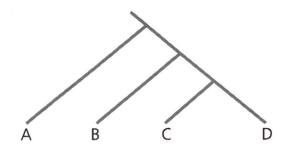

In this style, time is just the opposite of the figure above—the extant (living) species or groups are listed at the bottom, so each successive node up represents an earlier and earlier common ancestor.

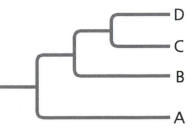

In this style, the extant (living) species or groups are listed at the right, so each successive node to the left represents an earlier and earlier common ancestor.

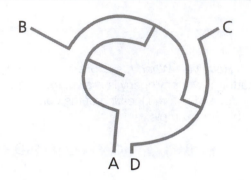

And in this style, time is depicted as concentric circles. Extant species are on the outside, and earlier and earlier common ancestors can be found closer and closer to the center.

Contemplate

What differences and similarities in DNA would you expect to find if our human ancestors interbred?	Why are conserved genes important when developing molecular phylogenies?

Delve Deeper

1. Do scientists use different lines of evidence to support and test phylogenetic hypotheses developed with molecular data?

2. How is the theory of neutral evolution different from the theory of evolution by natural selection? How is it the same?

3. As part of their analysis of Darwin's finches, Sato et al. (2001) developed this phylogeny for some of the tribes (a taxonomic unit that lies between family and genus) within the family Fringillidae (true finches) based on cytochrome b gene sequences. The tribes listed on the right are the currently accepted tribes based on previous DNA analyses. What does this tree tell you about the relationships within the Fringillids?

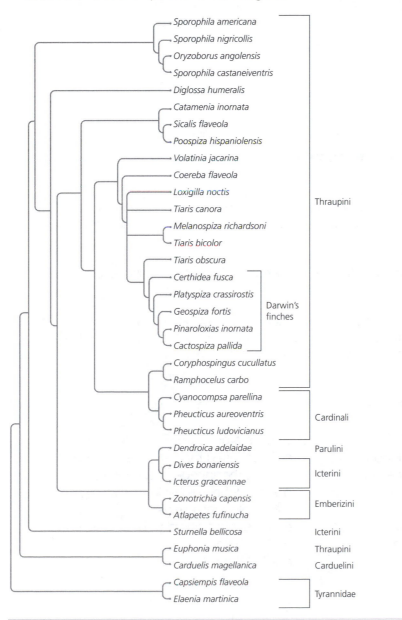

4. How can molecular data be used to understand natural selection in the past?

5. Chris Stringer proposed the hypothesis that all major ethnic groups of humans—Africans, Europeans, and Asians—were derived from recent African ancestry. How did Sarah Tishkoff and her colleagues test this hypothesis?

Go the Distance: Examine the Primary Literature

Wolfgang Enard, Molly Przeworski, Simon E. Fisher, Cecilia S. L. Lai, Victor Wiebe, Takashi Kitano, Anthony P. Monaco, and Svante Pääbo sequenced complementary DNAs that encode the *FOXP2* protein in a variety of species, including humans. They also investigated variation within the human FOXP2 gene.

- Why do they argue that natural selection has shaped the FOXP2 gene during recent human evolution?

Enard, W., M. Przeworski, S. E. Fisher, C. S. L. Lai, V. Wiebe, et al. 2002. Molecular Evolution of *FOXP2*, a Gene Involved in Speech and Language. *Nature* 418 (6900): 869–72. doi:10.1038/nature01025. http://www.nature.com/nature/journal/v418/n6900/abs/nature01025.html.

Test Yourself

1. What is meant by coalescence?
   a. The process of tracing homologous alleles back to a common ancestor.
   b. The history of an allele within a population.
   c. The point in the history of a genetic locus when it becomes polymorphic.
   d. All of the above.
   e. None of the above.

2. What is a gene tree?
   a. A tool used by molecular biologists to trace the lineage of a species.
   b. A branched genealogical lineage of homologous alleles.
   c. A graphic illustration of coalescence.
   d. Both b and c.
   e. All of the above.

3. Why don't all gene trees reflect the phylogeny of species?
   a. Because the branch lengths of a species tree are usually much longer on average than the coalescence times of the genes being analyzed.
   b. Because coalescence of specific genes can occur before speciation events.
   c. Because speciation events can sometimes be very rapid.
   d. Both b and c.
   e. All of the above.

4. Which of the following statements does Sarah Tishkoff's phylogenetic analysis of human mitochondrial DNA (Figure 9.10) illustrate?
   a. Humans that colonized India are no more distantly related to humans that colonized the Americas than humans that colonized the Middle East.
   b. Humans that colonized India, Central Asia, Europe, the Middle East, East Asia, the Americas, and Oceania represent a monophyletic clade.
   c. Humans that colonized the Americas share their most recent common ancestor with humans in East Asia.
   d. All of the above.
   e. None of the above.

5. Which of the following statements about the phylogenetic relationships among finches is TRUE?
   a. The Cocos finch shares a more recent common ancestor with Darwin's finches than with other finch species.
   b. The vegetarian finch is more closely related to the tree finches than to the ground finches.
   c. *Tiaris bicolor* is the common ancestor of all of Darwin's finches.
   d. *C. olivacea* is more closely related to *C. fusca* than to Darwin's finches.
   e. All of the above are true statements.

6. Molecular phylogenies indicate that
   a. The same mutation evolved in three separate lineages of HIV; in each instance

it improved the ability of the virus to infect humans.
   b. HIV came from a monkey virus that was introduced into people by contaminated vaccinations.
   c. HIV is a monophyletic strain of lentivirus that infects both humans and chimpanzees.
   d. The common ancestor of simian immunodeficiency virus and human immunodeficiency virus came from horses.
   e. None of the above.

7. The theory of neutral evolution describes
   a. The rate of mutation at a site that results from purifying selection, regardless of the size of the population.
   b. The rate of mutation at a site in the absence of selection, regardless of the size of the population.
   c. The competition between genetic drift and natural selection within the genome.
   d. Both a and b.
   e. Both b and c.

8. Which of these is a TRUE statement about molecular clocks?
   a. Molecular clocks use neutral theory to date events within a phylogeny.
   b. Molecular clocks can be calibrated using fossils of known age.
   c. Molecular clocks can be affected by the segments of DNA being examined and relative sizes of the populations.
   d. All of the above are true.
   e. None of the above is true.

9. What proportion of the human genome has no known function?
   a. 98.8 percent of our genome does not encode proteins and has no known function.
   b. Most of the 98.8 percent of our genome that does not encode proteins has no known function, but scientists are finding small, isolated segments that are important.

c. Scientists don't actually know what proportion of our genome has no known function because they are finding new methods of uncovering important elements.

d. 1.2 percent of our genome does not encode proteins and has no known function.

e. Every element of the human genome has a function.

10. Why do scientists consider *APOAV* a conserved gene?

a. Because it occurs so far away from other *APO* genes.

b. Because mice evolved before humans.

c. Because mice produce fewer triglycerides than humans.

d. Because of the high level of homology between mice and humans.

e. Because it encodes a protein important to lipid regulation.

Answers for Chapter 7
Check Your Understanding

1. Which of the following statements about genetic drift is TRUE?

a. Genetic drift occurs when populations move to new locations.

Incorrect. When a few members of a population colonize a new location, random processes such as genetic drift can have a great influence on the diversity of alleles. Genetic drift occurs in all populations, however, whether they move to new locations or not. Genetic drift describes the way allele frequencies "drift" randomly away from their starting values (see Chapter 6).

b. Because species need specific adaptations to survive, genetic drift cannot be an important factor in evolution.

Incorrect. Evolution cannot determine need—it is based on simple processes, such as random mutations that become more or less common over the course of generations. Genetic drift is also a random process. It is nonselective, and genetic drift can result in changes to a population that are not necessarily adaptive. Genetic drift may be an extremely important factor in evolution (see Chapter 6).

c. Genetic drift is a random process that contributes to evolutionary change within populations.

Correct. Genetic drift occurs in all populations, whether they are big or small or move to new locations or not. Genetic drift describes the way allele frequencies "drift" randomly away from their starting values. This process is an important contribution to evolutionary change (see Chapter 6).

d. Genetic drift is a form of natural selection.

Incorrect. Genetic drift is a nonselective process and not a form of natural selection. Like natural selection, however, genetic drift does influence the frequency of alleles in the next generation, but genetic drift can result in changes to a population that are not necessarily adaptive.

e. Genetic drift refers to the random mutations that affect small populations.

Incorrect. Genetic drift is a random process that affects allelic diversity, especially in small populations, because chance plays a role in the alleles that appear in the next generation. Mutations also occur randomly, and their persistence in a population results from the process of natural selection and genetic drift (see Chapter 6 for more on genetic drift).

2. Which of the follow are NOT noncoding regions of DNA?

a. Pseudogenes.

Incorrect. Pseudogenes are copies of what once was a functional gene, but mutations disabled them. Pseudogenes are not rare—a recent study estimated that the human genome includes 17,032 pseudogenes (see Chapter 5)!

b. Proteins.

Correct. Proteins are produced in protein-coding regions. Noncoding regions do not produce proteins by definition. Protein-coding genes make up only 1.2 percent of the human genome (see Chapter 5).

c. MicroRNAs.

Incorrect. MicroRNAs and other functional RNA molecules are noncoding regions in DNA. MicroRNAs function to regulate expression of other genes, including protein-coding regions (see Chapter 5).

d. Mobile genetic elements.

Incorrect. Mobile genetic elements are parasite-like segments of DNA. They make copies of themselves that are then reinserted into the genome, and as a result they make up over half of the human genome (see Chapter 5)!

3. Why is heritable variation among individuals an important factor for natural selection?

a. Because variation has to be inherited from parent to offspring for a species to survive.

Incorrect. Theoretically, variation does not have to be heritable for a species to continue in existence or for that matter for an individual to survive. In fact, horizontal gene transfer is an important mechanism of gene transfer that does not result from genes shared between parents and offspring. However, heritable variation is integral to natural selection for most organisms because it is an effective mechanism by which beneficial mutations may be transmitted. Natural selection acts to alter the abundances of those mutations in the population based on the relative reproductive success of individuals possessing those mutations (see Chapter 6).

b. Because variation has to be heritable for individuals to pass down their beneficial mutations.

Correct, but so are other answers. Heritable variation is integral to natural selection for most organisms because it is an effective mechanism by which beneficial mutations may be transmitted. Natural selection acts to alter the abundances of those mutations in the population based on the relative reproductive success of individuals possessing those mutations (see Chapter 6).

c. Because when individuals respond to the environment, they can pass that success on to offspring.

Incorrect. Jean-Baptiste Lamarck suggested that individuals that responded to the environment could pass those acquired characteristics to their offspring as a mechanism of evolution. If a phenotypic trait is not heritable, however, it can't be passed on to offspring. A parent who dyes his or her hair blond won't necessarily have blond children. Variation arises because of mutations, and in any given environment, some individuals do better than other individuals. When those mutations are heritable, natural selection acts to alter the abundances of those mutations in the population based on the relative reproductive success of individuals possessing those mutations (see Chapter 6).

d. Because natural selection cannot act when all individuals are absolutely identical.

Correct, but so are other answers. Individuals of a species or group must vary in some characteristic, and some of those characteristics must be "better" than others. Natural selection acts to alter the abundances of those characteristics in the population based on the relative reproductive success of individuals possessing them (see Chapter 6). So heritable variation is integral to natural selection for most organisms because it is an effective mechanism by which beneficial mutations may be transmitted.

e. Both b and d.

Correct. Individuals of a species or group must vary in some characteristic (e.g., via beneficial mutations), and some of those characteristics must be "better" than others. Natural selection acts to alter the abundances of those characteristics in the population based on the relative reproductive success of individuals possessing them (see Chapter 6). So heritable variation is integral to natural selection for most organisms because it is an effective mechanism by which beneficial mutations may be transmitted from one generation to the next.

Identify Key Terms

coalescence	The process by which, looking back through time, the genealogy of any pair of homologous alleles merges in a common ancestor.
conserved region	A functional segment of DNA that is almost identical among every species in which scientists have looked.
gene tree	The branched genealogical lineage of homologous alleles, which traces their ancestry back to an ancestral allele.
microsatellite	Noncoding stretches of DNA containing strings of short (1–6 base pairs), repeated segments. The number of repetitive segments can be highly polymorphic, so these segments serve as valuable genetic characters for comparing populations and for assigning relatedness among individuals (DNA fingerprinting).
molecular clock	A method used to determine time based on base pair substitutions. Rates of change can be used to deduce the divergence time between two lineages in a phylogeny, for example. Other markers of time, such as fossils with known ages, can be used to calibrate this methodology.

neutral theory of molecular evolution	A theory that describes the pattern of nucleotide sequence evolution under the forces of mutation and random genetic drift in the absence of selection. Under these conditions, the rate at which one nucleotide replaces another in a population (the substitution rate) should equal the rate of mutation at that site (mutation rate) regardless of the size of the population.
positive selection	Selection that increases the frequency of alleles within a population, such as mutations to non-synonymous sites that enhance essential functions.
purifying selection	Selection that removes deleterious alleles, such as mutations to non-synonymous sites that disrupt essential functions.
replacement substitution	Mutations that alter the amino acid sequence of a protein. These can affect the phenotype and are therefore more subject to selection.
species tree	The branched lineage of species, which traces the evolutionary history back to a common ancestor.
synonymous substitution	Mutations that do not alter the amino acid sequence of a protein. Because these mutations do not affect the protein an organism produces, they are less prone to selection and often free from selection completely.

Link Concepts

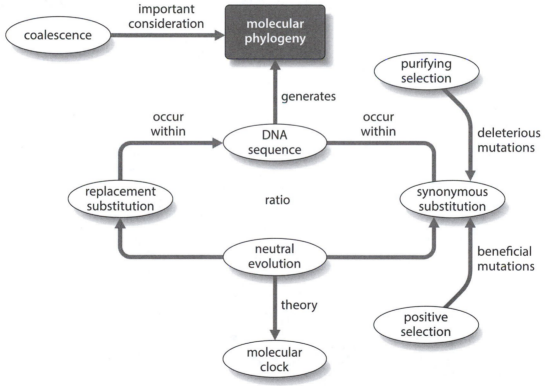

Interpret the Data

- What group shares the most recent common ancestor with the group of humans that colonized the Americas?
  Humans that colonized East Asia.
- According to Tishkoff et al. (2009), the longer branch lengths for groups of humans in the Americas, Oceania, and Pygmy, and some of the hunter-gatherers, indicate high levels of

genetic drift. Why might those groups have experienced higher levels of genetic drift than groups that remained in Africa?

Because groups that left Africa would likely have been small, and in small populations, alleles can become fixed due to chance alone.

Games and Exercises

	Horse	Donkey	Whale	Chicken	Penguin	Snake	Moth	Yeast	Wheat
Horse	0	1	5	11	13	21	29	46	46
Donkey		0	4	10	12	20	28	45	45
Whale			0	9	11	17	27	45	44
Chicken				0	3	17	28	47	46
Penguin					0	19	26	45	46
Snake						0	30	46	45
Moth							0	48	46
Yeast								0	48
Wheat									0

	Horse	Donkey	Whale	Chicken	Penguin	Snake	Moth	Yeast	Wheat
Human	12	11	10	13	13	14	29	44	44

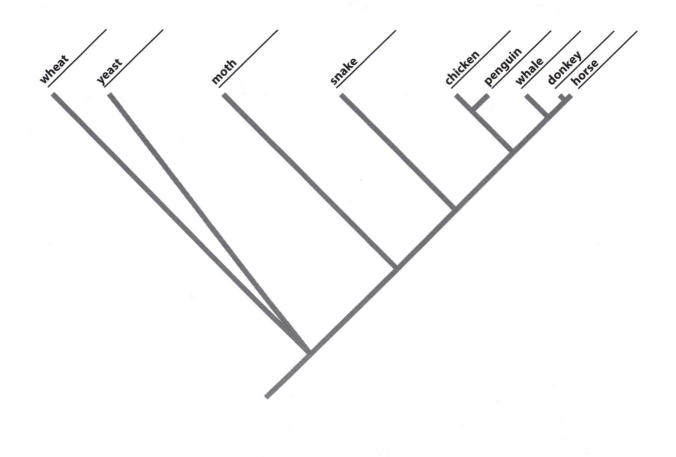

Delve Deeper

1. Do scientists use different lines of evidence to support and test phylogenetic hypotheses developed with molecular data?

   Yes. Scientists often check their results against different lines of evidence, such as the fossil record and morphological evidence. Molecular data can be used to develop phylogenetic hypotheses and tested with other lines of evidence, or molecular data can be used to support and/or test phylogenetic hypotheses developed with evidence from the fossil record or morphological data.

2. How is the theory of neutral evolution different from the theory of evolution by natural selection? How is it the same?

   The theory of neutral evolution addresses variation in nucleotide sequences and makes two predictions: that that variation is largely due to differences that are selectively "neutral," and that most evolutionary change is the result of genetic drift acting on neutral alleles. There are several ways that variation can be selectively neutral. Noncoding DNA, such as pseudogenes, is not likely to affect phenotypes, so mutations in those portions of the genome should not experience natural selection. Mutations to protein-coding genes may or may not affect the phenotype—the amino acid coding structure contains a lot of flexibility. For example, sequences of three nucleotides (codons) may differ and yet encode the same amino acid (CUU, CUC, CUA, CUG, UUA, and UUG all encode for leucine). So potentially, many single-nucleotide mutations are synonymous substitutions that are not expressed.

   Neutral theory also predicts that changes in the frequency of these neutral alleles will result from drift. Because genetic drift occurs when a random, nonrepresentative sample from a population produces the next generation, the frequencies of neutral alleles can change radically, being driven to fixation or extinction. This prediction leads to testable hypotheses about natural selection because scientists can compare the substitutions that occur in replacement sites to the substitutions that occur in synonymous sites.

   The theory of evolution by natural selection requires that some aspect of the phenotype be affected by mutation, so any single-nucleotide mutation that is a replacement substitution and is expressed will lead to heritable variation among individuals. The sample producing the next generation is representative of the fitness of alleles, and change in allele frequency is the outcome.

   Like natural selection, neutral evolution in sexually reproducing organisms requires mutations to affect germ lines in order to be heritable (in plants, somatic mutations may be transferred to offspring). Both the theories of neutral evolution and evolution by natural selection explain variation among individuals, and both predict that allele frequencies will change over time.

3. As part of their analysis of Darwin's finches, Sato et al. (2001) developed this phylogeny for some of the tribes (a taxonomic unit that lies between family and genus) within the family Fringillidae (true finches) based on cytochrome b gene sequences. The tribes listed on the right are the currently accepted tribes based on previous DNA analyses. What does this tree tell you about the relationships within the Fringillids? (Darwin's Finches = DF in the figure.)

   The tree indicates that Darwin's finches share a common ancestor with *Tiaris obscura*, but the tribes developed in previous analyses may need to be reconsidered. For example, Darwin's finches are wholly within the tribe Thraupini, and the old tribes don't match up with the clades developed using cytochrome b. One species formally classified as a Thraupini (*Euphonia musica*) is actually distantly related. Combined with several other inconsistencies, this phylogeny indicates that the traditional tribal taxa may need to be revised.

4. How can molecular data be used to understand natural selection in the past?

   Scientists can look for evidence of different kinds of selection in the past by comparing the ratios of synonymous to replacement substitutions within coding regions. The probability of a substitution at a synonymous site should be equal to the probability of a substitution at a replacement site. When they are not equal, scientists can reject the hypothesis of neutral evolution. Then they can look at the patterns of the ratios of synonymous to replacement substitutions. For example, strong positive selection on a gene can lead to unusually large numbers of replacement substitutions that change the structure of proteins, whereas purifying selection can eliminate replacement mutations to genes with essential functions that are easily disrupted by mutations, leading to a very low number of replacement substitutions compared to synonymous ones. For example, the *FOXP2* gene has changed very little in our common ancestors, but in humans, two amino acids have changed in the protein in the past 6 million years and may contribute to our human-ness.

5. Chris Stringer proposed the hypothesis that all major ethnic groups of humans—Africans, Europeans, and Asians—were derived from recent African ancestry. How did Sarah Tishkoff and her colleagues test this hypothesis?

Stringer's hypothesis makes a prediction: if the major ethnic groups of humans shared a common ancestor, Europeans and Asians should share more derived characters with each other than with Africans. The alternative, or null hypothesis, would make no such prediction. Tishkoff and her colleagues examined more than 1300 genetic loci for patterns of variation. Based on the quantity of that variation, they used a neighbor joining distance matrix to develop a phylogenetic tree. They found that all non-Africans form a monophyletic group, indicating that non-Africans share more derived characters with each other than with Africans. In addition, they found that genetic variation was greater in Africans than in these non-African groups. This pattern is typical of a founder event (a type of genetic drift that accompanies the founding of a new population from a very small number of individuals). These results support the hypothesis that non-Africans are descendants of a common ancestor shared with Africans, one that likely migrated out of Africa.

Test Yourself

1. d; 2. d; 3. d; 4. d; 5. a; 6. a; 7. b; 8. d; 9. c; 10. d

Chapter 8
ADAPTATION: THE BIRTH OF THE NEW

Check Your Understanding

1. Which of the following support(s) the concept of common descent?
   a. All living organisms share a common code of genetic instructions.
   b. Flowers and insects often share a helpful partnership.
   c. The regularity of homologies among species (e.g., mammals have the same arrangement of bones in their forelimbs).
   d. Both a and c.
   e. All of the above.

2. Which of the following statements about genetically controlled traits is TRUE?
   a. Interactions among the alleles at a variety of genetic loci can affect the expression of a trait, such as height, and these interactions can differ from individual to individual.
   b. A single genotype may produce different phenotypes depending on the environment.
   c. A single mutation to a regulatory gene can affect many phenotypic traits.
   d. All of the above.
   e. a and b only.

3. Which of the following mutations does NOT affect the expression of a gene?
   a. Mutations to promoter regions.
   b. Mutations that are synonymous.
   c. Mutations to regulatory elements.
   d. Mutations that code for hormones.

Learning Objectives for Chapter 8:

➢ Explain how gene duplication can lead to new adaptations.
➢ Describe two examples where proteins were co-opted for other functions.
➢ Explain why *Hox* genes are considered part of the "genetic toolkit."
➢ Explain how changes in the timing and location of the expression of developmental genes can affect the appearance of important traits like legs.
➢ Describe three important steps in the evolution of the vertebrate eye.
➢ Distinguish the outcomes between a mutation in pleiotropic gene and a non-pleiotropic gene.
➢ Analyze the imperfections of a familiar complex adaptation.
➢ Compare and contrast parallelism and convergent evolution.

The Tangled Bank Study Guide

Identify Key Terms

Connect the following terms with their definitions on the right:

anterior	Mutations, characters, or traits that appeared long before they were co-opted for a use subject to natural selection. (An alternative definition addresses characters or traits that were shaped by natural selection for one function but co-opted for a new use.)
complex adaptation	Genes that are similar in sequence and function because they were inherited from a common ancestor.
convergent evolution	Suites of coexpressed traits that together experience selection for a common function. These include phenotypes that are influenced by many environmental and genetic factors and when multiple components must be expressed together for the trait to function or as novel traits that require multiple mutations to achieve a fitness advantage.
crystallins	A mutation that causes a segment of DNA to be copied a second time. The mutation can affect a region inside a gene, an entire gene, and in some cases, an entire genome.
dorsal	A term generally referring to the back side of an organism (not its rear). The reference may vary depending on the organism.
evo-devo	Relatively short sequences of DNA (approximately 180 base pairs) characteristic of some of the genes that control the development of organisms.
exaptation	Similar traits that evolve in unrelated lineages. Examining traits that converge can shed light on how each lineage arrived at the shared phenotype—the genetic and developmental underpinnings of similar adaptations. It can reveal multiple solutions to a common challenge as well as the limitations imposed by history within each lineage.
gene duplication	A group of light-sensitive proteins that were likely present very early in the evolution of animals and began to diversify in the common ancestors of bilaterians and cnidarians.
gene expression	The co-option of a particular gene for a totally different function as a result of a

126 | CHAPTER 8: Adaptation

	mutation. The reorganization of a preexisting regulatory network can be a major evolutionary event.
gene recruitment	A group of proteins recruited to focus light in the eye. Other proteins in the group have diverse functions in the body, such as heat-shock proteins.
genetic toolkit	A type of "patterning" gene important in demarcating the geography of developing animals, determining the relative locations and sizes of body parts. These patterning genes are part of the "genetic toolkit" inherited by all animals with bilateral body symmetry.
homeobox	The process by which information from a gene is transformed into a product.
homologous genes	Convergent evolution that results from mutations to the same gene in different lineages.
Hox gene	A term generally referring to the front side of an organism. The reference may vary depending on the organism.
opsins	The condition when a mutation in a single gene affects the expression of many different phenotypic traits. If a mutation with beneficial effects for one trait also causes detrimental effects on other traits, the effect is considered antagonistic.
parallelism	A term generally indicating regions toward the tail of an organism. The reference may vary depending on the organism.
pleiotropy	A term generally indicating regions toward the head of an organism. The reference may vary depending on the organism.
posterior	A field of evolutionary biology that examines changes in development of organisms over time.
ventral	A subset of the genes in an organism's genome that control development. The genes within the subset are highly conserved across organisms.

Link Concepts

1. Fill in the following concept map with the key terms from the chapter:

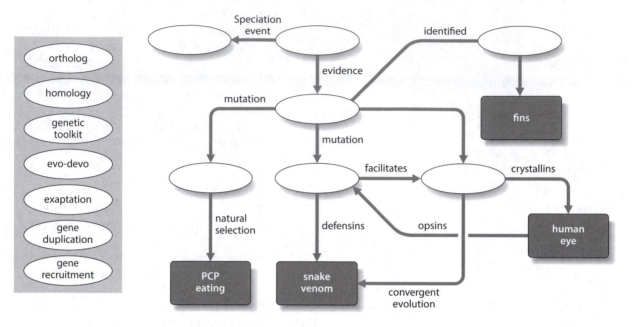

2. Develop your own concept map that explains how evolution imposes constraints on adaptation. Think about mutations, pleiotropy, genetic toolkits, homology, body-patterning gene networks, and lineages. Do you think placental and marsupial mammals faced different constraints in the course of their evolution?

Interpret the Data:

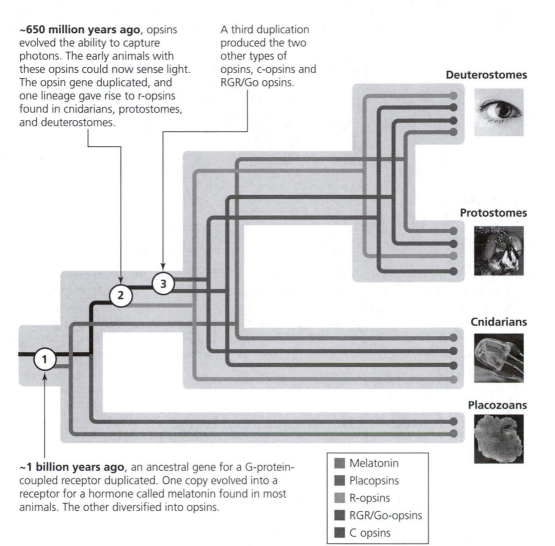

~650 million years ago, opsins evolved the ability to capture photons. The early animals with these opsins could now sense light. The opsin gene duplicated, and one lineage gave rise to r-opsins found in cnidarians, protostomes, and deuterostomes.

A third duplication produced the two other types of opsins, c-opsins and RGR/Go opsins.

Deuterostomes

Protostomes

Cnidarians

Placozoans

~1 billion years ago, an ancestral gene for a G-protein-coupled receptor duplicated. One copy evolved into a receptor for a hormone called melatonin found in most animals. The other diversified into opsins.

- ■ Melatonin
- ■ Placopsins
- ■ R-opsins
- ■ RGR/Go-opsins
- ■ C opsins

Gene duplication events were crucial to the evolution and diversification of opsins so important for human vision. Scientists have been able to trace the origins of opsins by examining similar proteins across taxa. Opsins evolved from a family of proteins known as G-protein-coupled receptors (GPCRs). These GCPRs are found in all animals (and even fungi).

Are all opsins used in vision?

How many types of opsins can be found in the deuterostomes (a group of animals that includes humans)?

Are cnidarians sensitive to light? Support your argument.

What mutation was necessary for the evolution of opsins? Why was that important?

Games and Exercises

Paint by Numbers

As scientists began to explore genome sequences in various model organisms, they discovered amazing patterns! For example, in genes that influence the shapes of our bodies, like *Hox* genes, they discovered segments about 180 bases long that were virtually identical across organisms. They named these segments homeoboxes, and these homeoboxes translate into protein sequences that are 60 amino acids in length. These regulatory genes act as switches, controlling suites of other genes in cascades that ultimately influence how we develop. These homeobox genes are strong evidence for the shared ancestry among bilateral organisms.

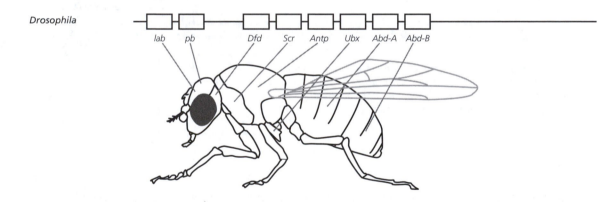

Seeing some of the homologies scientists have discovered is as easy as paint by numbers. In the following diagram, the genes are lined up, literally, from head to tail on the chromosomes. Start with the *Drosophila* genes: labial (*lab*), proboscipedia (*pb*), deformed (*Dfd*), sex combs reduced (*Scr*), antennapedia (Antp), ultrabithorax (Ubx), and abdominal-A (Abd-A). Simply choose different colors for each of the boxes (genes) indicated on the *Drosophila* chromosome and color each one. Color the homologous boxes for the Hox genes on the four human chromosomes with the corresponding colors from the Drosophila chromosome.

Match these important developmental genes with the areas where they are expressed in the *Drosophila* larva, the *Drosophila* adult, the mouse larva, and the human larva. Color each "domain of expression" with the color corresponding to the genes.

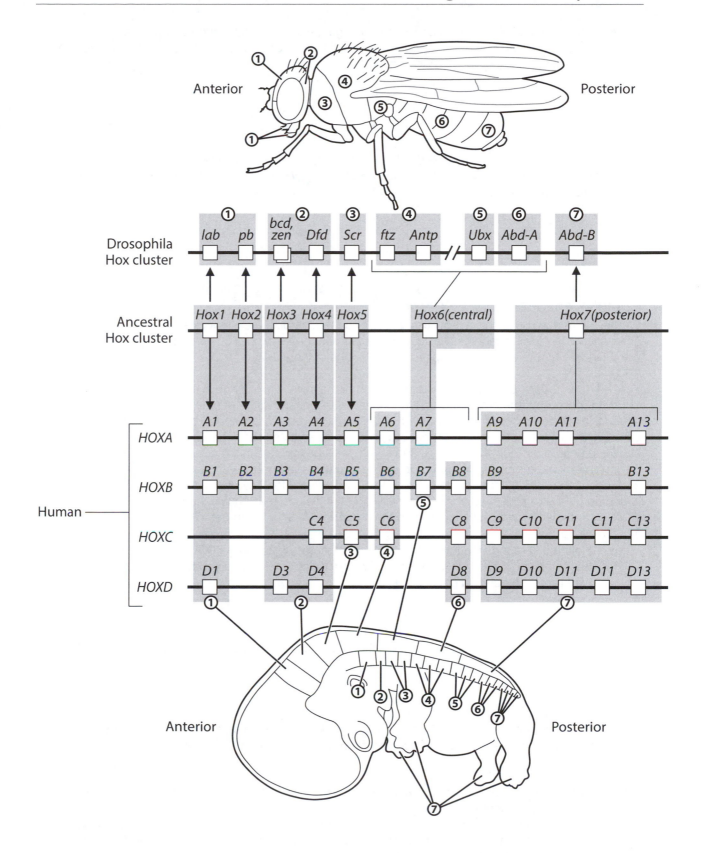

- The diagram indicates that *Drosophila Hox* genes lie on a single chromosome in a single cluster, and mouse and human *Hox* genes lie in clusters on four different chromosomes. What kind of mutation(s) likely produced this diversity?

- Hox genes are also expressed in the limbs of Stage 19 human embryos (specifically, 5'*HoxA* and *HoxD* genes). Which ortholog found in *Drosophila* gave rise to *Hox* genes governing human limb development?

Complex Adaptations and Selection

Werner Heim developed this activity using regular playing cards to help illustrate how complex adaptations can arise given cumulative, nonrandom natural selection. You can do this activity alone or with friends, but make sure to complete both versions of the activity.

You will need a deck of cards. Separate out each of the suits, choose one suit, and shuffle it thoroughly. Shuffling represents the probability of a functional mutation, and each round represents time. So you'll need to tally the number of times you shuffle the cards on a piece of paper. For round 1, shuffle the cards, then look at the top card. If the top card is the ace, you can start an "organism" stack. If not, shuffle the cards for round 2, round 3, and so on, until the top card is the ace. Once you've started the "organism" stack, the objective is to build up the mutations through the entire suit of cards by shuffling and recording rounds. If the top card is the next card needed for the construction of the organism, place it face up on the organism stack, shuffle, and repeat. If it's not, then shuffle the cards again. Continue shuffling and placing appropriate cards on the organism stack until you get to the king.

- How many rounds did it take to complete the complex adaptation? What do you think would happen if you repeated the experiment? How many times would you have to repeat the experiment to get the right answer?

- How does this simulation compare with the cumulative mutations necessary for the evolution of a complex adaptation?

- Compare the evolution of citrate feeding with being forced to start with the ace? What might the outcome be in terms of citrate feeding if the ace wasn't the first mutation?

Now repeat the procedure, but this time, there is no "organism" stack—you don't get to work with cumulative mutations. Shuffle the cards thoroughly. Keep track of the number of rounds, just like in the previous experiment. For the first round, examine all of the cards. Are they in order from ace to king? If not, shuffle the cards and examine the cards again. Continue shuffling and examining until the order of cards comes out in order from ace to king.

- How many rounds did it take to complete the complex adaptation? How many rounds did you try? How long do you think it would take to complete the complex adaptation simply by shuffling cards?

Adapted from Heim, Werner G. 2002. Natural Selection among Playing Cards. *American Biology Teacher* 64 (4): 276–78. doi:http://dx.doi.org/10.1662/0002-7685(2002)064[0276:NSALC]2.0.CO;2

 also available from the ENSI websitehttp://www.indiana.edu/~ensiweb/lessons/ns.cum.l.html

Go Online

 Let's Shake Wings

Alexander Vargas and John Fallon examined development in chicken embryos and found evidence that may clarify the relationship between dinosaurs and birds.

http://news.sciencemag.org/sciencenow/2005/01/11-01.html?ref=hp

Quantity Is Key for Hox *Genes*

 Hox proteins apparently aren't restricted to single tasks—they may actually be able to switch places fairly easily. The amount of proteins made, however, may be the key to developmental patterns.

http://news.sciencemag.org/sciencenow/2000/02/14-02.html?ref=hp

Evolutionary Development: Chicken Teeth—Crash Course Biology #17

 In this Crash Course video, Hank Green introduces evo-devo, *Hox* genes, and the process of development.

http://www.youtube.com/watch?v=9sjwlxQ_6LI

Evolution: Great Transformations

PBS *Evolution* shows how scientists discovered the genetic toolkit. Scientists were able to switch the patterning genes for eyes between fruit flies and mice and show that not only did the two species use the same mechanism for gene expression, they also used the same gene.

http://www.pbs.org/wgbh/evolution/library/03/4/l_034_04.html

Evolution: Darwin's Dangerous Idea

Zoologist Dan-Erik Nilsson developed a model for the evolution of the eye. This video segment from PBS *Evolution* shows the important evolutionary steps that transformed eyes and the organisms that possess functional eyes at these various stages.

http://www.pbs.org/wgbh/evolution/library/01/1/l_011_01.html

Common Misconceptions

Different Cell Types of an Individual Carry the Same DNA

If a cell is a diploid, nucleated cell, it contains all the organism's chromosomes, half from the female and half from the male. It doesn't matter whether those cells are kidney cells, muscle cells, brain cells, or skin cells. Therefore, the genes within every single diploid, nucleated cell can be regulated, whether that regulation occurs as functions within the cell or outside of it. In fact, the differences among cell types arise primarily because different sets of genes are expressed in different cell types. The DNA is not different.

Complexity in Evolution Is Explainable

Complex adaptations, like the human eye, can be explained by evolutionary theory. Many different lines of evidence (physiology, chemistry, genetics, phylogenetics, and such) show that mutations, gene duplication, and gene recruitment are all processes that can combine to produce incredible structures. Although the mutations may be random—a matter of chance—their persistence in a population is definitely **not** random if the mutation affects the reproductive success of the individuals that have it. And *any* functional advance can affect the fitness of organisms—it doesn't have to be the *pinnacle of all complex adaptations* to be effective. As humans, we see human eyes as the best of all eye adaptations, but there are many different solutions to the problem of detecting light. Numerous organisms have different versions of light-sensing organs, from very simple to very complex, and some are more sophisticated and more sensitive than ours! These adaptations are all the result of many small changes over time, each small change adding a new level of functionality.

Contemplate

How might snakes have evolved from a common ancestor with legs?	How quickly—evolutionarily speaking—could a genetic toolkit be co-opted to produce a novel trait?

Delve Deeper

1. What are the differences and similarities between gene duplication and gene recruitment?

2. Why do mice and flies have the same genetic toolkit of patterning genes?

3. How was a defensin gene co-opted for predation in snakes?

4. Why are insects constrained from growing any larger in their current environment?

5. Describe the role of opsins and crystallins in the evolution of the vertebrate eye.

6. Consider similar traits in both sharks and dolphins. What evidence would you use to identify a trait as being homologous and a different trait as being convergent?

7. Are developmental genetic regulatory networks perfectly designed? Explain why or why not.

Go the Distance: Examine the Primary Literature

Irma Varela-Lasheras and her colleagues continue work on the evolutionary constraints imposed by changes in the number of cervical vertebrae. They examine sloths and manatees, two mammal species that are exceptions to the seven-vertebrae rule—sloths with eight cervical vertebrae and manatees with six.

- What are homeotic transformations?

- What five predictions did the authors make about homeotic transformations?

- According to the authors, how can the pleiotropic constraints on body plan changes, including changes to the number of cervical vertebrae in mammals, be overcome?

- Why might this discovery be important?

Varela-Lasheras, I., A. Bakker, S. van der Mije, J. Metz, J. van Alphen, et al. 2011. Breaking Evolutionary and Pleiotropic Constraints in Mammals: On Sloths, Manatees, and Homeotic Mutations. *EvoDevo* 2 (1): 11. doi:10.1186/2041-9139-2-11. http://www.evodevojournal.com/content/2/1/11.

Test Yourself

1. How do novel traits arise?
   a. Through an existing trait that reverts to a previous version.
   b. Through mutation leading to new phenotypes.
   c. They are inherited from the previous generation.
   d. Through a new combination of traits already existing in the individual.

2. Which of the following statements about Richard Lenski's *E. coli* populations is false according to the following figure?

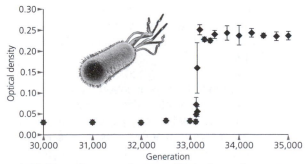

a. Initially, only a small proportion of one lineage of *E. coli* could feed on citrate.

b. Citrate feeding did not evolve in the lineage for 30,000 generations.

c. The citrate feeding trait quickly came to dominate the *E. coli* lineage.

d. One generation of *E. coli* showed significant variation in the citrate feeding trait.

e. The frequency of citrate feeding was highly variable in all generations within the lineage of *E. coli*.

3. Which process led to the evolution of snake venom?
   a. Gene recruitment.
   b. Natural selection.
   c. Gene duplication.
   d. All of the above.
   e. None of the above.

4. Do scientists have evidence to support their hypotheses about the early evolution of eyes?
   a. No. Scientists have only a hypothesis based on the structure of proteins.
   b. No. The evidence for the evolution of eyes is inconsistent among lineages, such as bilaterians and cnidarians.
   c. Yes. Scientists know exactly how the eye evolved based on a series of steps involving mutations of different proteins.
   d. Yes. Scientists use evidence from phylogenies based on shared derived morphological structures as well as gene sequences and entire genomes.

5. Which organism does NOT use crystallins in its eyes to focus light?
   a. Humans.
   b. Octopuses.
   c. Fish.
   d. Ragworms.

6. How did scientists test the hypothesis that insect size is constrained by access to oxygen?
   a. They looked at insects of different sizes and compared how well they survived and reproduced.
   b. They controlled oxygen concentrations and determined how large insects could grow.
   c. They compared the sizes of insects in the fossil record when oxygen concentrations were higher to when oxygen concentrations were lower.
   d. They examined the fossil record of large insects and then looked for a factor that provided some explanation.

e. They compared the sizes of the tubes insects use to get oxygen between large and small fossil insects.

7. Why do giraffes have a recurrent laryngeal nerve that is 19 feet longer than it needs to be?
   a. Their ancestors were fish and did not have a larynx.
   b. The route of that nerve was inherited from their ancestors.
   c. The nerve makes more connections to their lungs than in their ancestors.
   d. The route that the nerve takes is the result of random processes.

8. Why are the wolf and the Tasmanian wolf (Figure 8.18) considered examples of convergent evolution?
   a. Because even though they are from two different lineages of mammals, their genotypes converged on the same form and ecological function.
   b. Because even though they were found on different continents, they have identical phenotypes.
   c. Because they share a common ancestor.
   d. Because their phenotypes converged on similar forms and ecological functions even though they are from two different lineages of mammals.
   e. Because their adaptations arise from similar developmental underpinnings.

9. How is parallelism different from convergent evolution?
   a. Parallelism occurs within the same lineage, and convergent evolution occurs in different lineages.
   b. Parallelism is the evolution of similar adaptations by closely related groups, and convergent evolution occurs within a single lineage.
   c. It's not different. Parallelism is convergent evolution that occurs because of similar mutations that arise independently in the same genes in different lineages.
   d. Parallelism is the evolution of similar adaptations by groups separated in time, and convergent evolution is the evolution of similar adaptations by groups separated geographically.
   e. Parallelism leads to identical genotypes, and convergent evolution leads to identical phenotypes.

10. What do you predict would happen if you inserted and expressed a Pax-6 gene from a mouse into the embryo of an octopus?
    a. No eyes would develop in the octopus.
    b. The normal two eyes would develop, but they would not work.
    c. Extra eyes would develop, but with mouse crystallins.
    d. Extra eyes would develop, but with octopus crystallins.

Answers for Chapter 8
Check Your Understanding

1. Which of the following support(s) the concept of common descent?

   a. All living organisms share a common code of genetic instructions.

   Correct, but so are other answers. The fact that all living organisms share a common code of genetic instructions is powerful evidence that all organisms share a common ancestor. This genetic code may seem simple, using only four nucleobases, but scientists are beginning to understand that the genome is far more complex than simple sequences of bases that code for amino acids. The genome also includes regulatory elements, pseudogenes, and genes encoding functional RNA molecules that along with hormones and other gene products interact to influence the phenotype of organisms (see Chapter 5). In fact, as you'll learn in Chapter 8, not only do organisms share a common code, they also share similar genes that are so structurally conserved that they perform similar tasks and can actually be experimentally interchanged among organisms without effect. This shared architecture provides additional evidence for common descent.

   b. Flowers and insects often share a helpful partnership.

   Incorrect. Although many flowers share a helpful partnership with insects, these relationships do not provide evidence of a shared common ancestor. Flowers and insects do not share recent common ancestors—their lineages split a long time ago. Organisms may coevolve, with members of each group acting as agents of selection on the other, and this coevolution can shape lineages. However, helpful partnerships in and of themselves do not necessarily support common descent.

   c. The regularity of homologies among species (e.g., mammals have the same arrangement of bones in their forelimbs).

   Correct, but so are other answers. The frequency of organisms that share homologies is strong evidence for common descent (see Chapter 4). Comparative biologists have been examining homologies in the fossil record for 150 years, and they've used these structures to organize and classify organisms into groups based on similarities, and more importantly, shared derived similarities. The most likely explanation for this organization is that homologies exist because the organisms shared a common ancestor. More recently, scientists have been examining homologies within genes and genomes (see Chapter 7), and this evidence is both confirming and changing the way we think about how organisms are related to each other.

   d. Both a and c.

   Correct. The fact that all living organisms share a common code of genetic instructions (Chapter 5) and that organisms frequently share homologies (Chapter 4) are both powerful evidence for common descent. In fact, as you'll learn in Chapter 8, not only do organisms share a common code, they also share similar genes that are so structurally conserved that they perform similar tasks and can actually be experimentally interchanged among organisms without effect. This shared architecture provides additional evidence for common descent.

   e. All of the above.

   Incorrect. Although the fact that all living organisms share a common code of genetic instructions (Chapter 5) and that organisms frequently share homologies (Chapter 4) are both powerful evidence for common descent, helpful partnerships in and of themselves do not serve as evidence that lineages share a common ancestor.

2. Which of the following statements about genetically controlled traits is TRUE?

   a. Interactions among the alleles at a variety of genetic loci can affect the expression of a trait, such as height, and these interactions can differ from individual to individual.

   Correct, but so are other answers. Many traits are polygenic—they are influenced by many genetic loci. The expression of human height, for example, may depend on interactions between alleles at different loci, and/or it may result from interactions between alleles and the environment (e.g., phenotypic plasticity). So height can vary among individuals because individuals have different alleles for different genes, and these alleles can vary in their contributions to height because of the variation in the potential interactions among the genetic loci and the environment (see Chapter 5).

b. A single genotype may produce different phenotypes depending on the environment.
Correct, but so are other answers. Phenotypic variation among individuals can result when a single genotype responds to different environments. The environment can be internal (such as regulatory genes and noncoding elements), or external (such as light or temperature). A plant grown in low light may have large leaves and small roots and in high light just the opposite, but that is not necessarily evolution. The evolution of phenotypic plasticity comes when a population, such as plants, becomes more or less sensitive to light as a cue to grow different sized leaves, for example (see Figure 5.13).

c. A single mutation to a regulatory gene can affect many phenotypic traits.
Correct, but so are other answers. Traits are often highly interconnected and complex. Regulatory genes, for example, often influence the expression of more than just a single gene—they influence many genes and therefore potentially many phenotypic traits. A mutation to this kind of gene is likely to be pleiotropic (affecting many traits), and a single base change can have varied consequences (see Chapter 5).

d. All of the above.
Correct. Many traits are polygenic—they are influenced by many genetic loci. Individuals can vary because they have different alleles for different genes, and these alleles occur in different combinations in different individuals, resulting in diverse effects from individual to individual. Genotype x environment interactions can contribute to variation when alleles interact with the environment in ways that affect the expression of a phenotype. Finally, a single mutation can affect many phenotypic traits if, for example, the mutation occurs in a regulatory gene (see Chapter 5).

e. a and b only.
Incorrect. Many traits are polygenic—they are influenced by many genetic loci—and those traits can vary depending on genotype x environment interactions. The expression of human height, for example, may depend on interactions between alleles at different loci, and/or it may result from the interaction of the

different alleles with the environment. So height can vary among individuals because individuals have different alleles for different genes, and these alleles can vary in their contributions to height because of the variation in the potential interactions among the genetic loci. Plus, mutations to some genes can have cascading effects, especially if that mutation affects a regulatory gene. A mutation to a gene that affects the ability of *another* gene to produce proteins, either at all or for any length of time, can influence the development of an organism tremendously (Chapter 5).

3. Which of the following mutations does NOT affect the expression of a gene?
a. Mutations to promoter regions.
Incorrect. Promoter regions are regions near one end of a gene that initiate transcription (Chapter 5). Mutations to these regions can affect whether a gene is expressed or not, or whether it is only partially expressed. In other words, a mutation to a promoter region can determine whether an important protein, like growth hormone, is produced or not.

b. Mutations that are synonymous.
Correct. Synonymous substitutions are mutations that do not affect the amino acid being encoded. Some amino acids are encoded by more than one codon (a triplet of bases; see Figure 5.5), so a mutation that substitutes one base for another without affecting the amino acid doesn't affect the expression of that gene. Figure 5.5 shows the codons for the 20 amino acids used to generate proteins. Note the diversity: some amino acids have many opportunities for synonymous substitutions—they are encoded by a number of different codons—and some have very few opportunities.

c. Mutations to regulatory elements.
Incorrect. Binding sites for regulatory elements (promoters) are found in a gene's control region. A gene's control region can have a number of regulatory regions, each influencing where different molecules can bind. The regulatory molecules bind to the DNA and change its shape, ultimately affecting how the gene is expressed. A mutation to one of these elements can affect not only the shape of the DNA but also whether that gene is expressed at all or for how long (Chapter 5).

d. Mutations that code for hormones.

Incorrect. Hormones are chemicals released by the body that function as signals. For example, adrenalin is a hormone produced in the adrenal gland near the kidneys. As adrenalin circulates around the body, it attaches to the surface of muscles and other types of cells, switching on other genes inside those cells. A mutation that affects the production of adrenalin can affect the amount of time muscle cells are exposed to adrenalin, which in turn can affect how long the genes inside are turned on or off—possibly affecting how long an individual could run away from a scary situation. If the mutation alters the hormone itself, the animal may not have a fight or flight reaction at all. So mutations that affect the production of hormones can have significant effects on gene expression.

Identify Key Terms

anterior	A term generally indicating regions toward the head of an organism. The reference may vary depending on the organism.
complex adaptation	Suites of coexpressed traits that together experience selection for a common function. These include phenotypes that are influenced by many environmental and genetic factors, and when multiple components must be expressed together for the trait to function or as novel traits that require multiple mutations to achieve a fitness advantage.
convergent evolution	Similar traits that evolve in unrelated lineages. Examining traits that converge can shed light on how each lineage arrived at the shared phenotype—the genetic and developmental underpinnings of similar adaptations. It can reveal multiple solutions to a common challenge as well as the limitations imposed by history within each lineage.
crystallins	A group of proteins recruited to focus light in the eye. Other proteins in the group have diverse functions in the body, such as heat-shock proteins.
dorsal	A term generally referring to the back side of an organism (not its rear). The reference may vary depending on the organism.
evo-devo	A field of evolutionary biology that examines changes in development of organisms over time.
exaptation	Mutations, characters, or traits that appeared long before they were co-opted for a use subject to natural selection. (An alternative definition addresses characters or traits that were shaped by natural selection for one function but co-opted for a new use.)
gene duplication	A mutation that causes a segment of DNA to be copied a second time. The mutation can affect a region inside a gene, an entire gene, and in some cases, an entire genome.
gene expression	The process by which information from a gene is transformed into a product.
gene recruitment	The co-option of a particular gene for a totally different function as a result of a mutation. The reorganization of a preexisting regulatory network can be a major evolutionary event.
genetic toolkit	A subset of the genes in an organism's genome that control development. The genes within the subset are highly conserved across organisms.
homeobox	Relatively short sequences of DNA (approximately 180 base pairs) characteristic of some of the genes that control the development of organisms.
homologous genes	Genes that are similar in sequence and function because they were inherited from a common ancestor.

Hox gene	A type of "patterning" gene important in demarcating the geography of developing animals, determining the relative locations and sizes of body parts. These patterning genes are part of the "genetic toolkit" inherited by all animals with bilateral body symmetry.
opsins	A group of light-sensitive proteins that were likely present very early in the evolution of animals and began to diversify in the common ancestors of bilaterians and cnidarians.
parallelism	Convergent evolution that results from mutations to the same gene in different lineages.
pleiotropy	The condition when a mutation in a single gene affects the expression of many different phenotypic traits. If a mutation with beneficial effects for one trait also causes detrimental effects on other traits, the effect is considered antagonistic.
posterior	A term generally indicating regions toward the tail of an organism. The reference may vary depending on the organism.
ventral	A term generally referring to the front side of an organism. The reference may vary depending on the organism.

Link Concepts

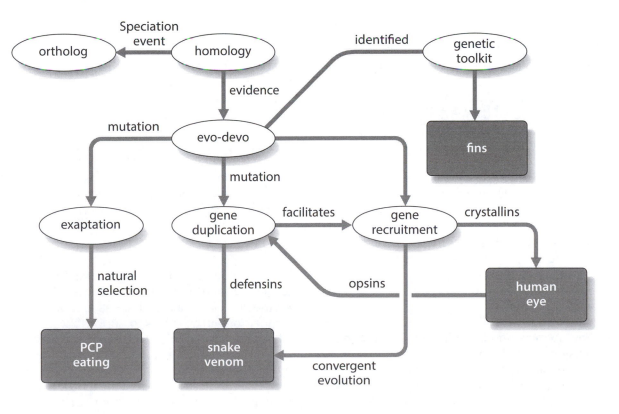

Interpret the Data

- Are all opsins used in vision?
  Not necessarily. Placopsins branched off before opsins could sense light.
- How many types of opsins can be found in the deuterostomes (a group of animals that includes humans)?
  Three: R-opsins, RGR/Go-opsins, and C-opsins.
- Are cnidarians sensitive to light?
  Yes. They have receptors for R-opsins, C-opsins, and RGR/Go-opsins.
- What mutation was necessary for the evolution of opsins? Why was that important?
  Duplication of the opsin gene. Duplication was important because the original gene could still function to produce some receptors while evolution acted on the duplicated gene to produce other receptors.

Games and Exercises

- The diagram indicates that *Drosophila Hox* genes lie on a single chromosome in a single cluster, and mouse and human *Hox* genes lie in clusters on four different chromosomes. What kind of mutation(s) likely produced this diversity?
  Duplication events that allowed diversification over the course of time.
- *Hox* genes are also expressed in the limbs of Stage 19 human embryos (specifically, 5'*HoxA* and *HoxD* genes). Which ortholog found in *Drosophila* gave rise to *Hox* genes governing human limb development?
  Abdominal-B (Abd-B).

Delve Deeper

1. What are the differences and similarities between gene duplication and gene recruitment?
   Gene duplication involves the copying of an existing gene, which may allow one of the copies to take on a new role. Gene recruitment occurs when a gene or gene network is transformed through mutation to perform a totally different function. Both can involve genes switching roles, but they differ in the source of the gene and how similar or different the final role will be from its original function. Gene duplication events often facilitate subsequent gene recruitment because the duplicated genes are no longer constrained by selection to perform their original function.

2. Why do mice and flies have the same genetic toolkit of patterning genes?
   Both mice and flies have bilateral body symmetry, and bilateral symmetry is an ancestral trait. The basic genetic toolkit for bilateral symmetry was inherited from the ancestor of all bilaterians.

3. How was a defensin gene co-opted for predation in snakes?
   Defensin genes, used to produce proteins that fight pathogens, were originally expressed in the pancreas of snakes and their ancestors. These genes began to be expressed in the mouth of the snakes as the result of a regulation mutation. Further mutations to the protein product allowed it to break down the muscles of the snake's prey. And finally, additional mutations combined with natural selection increased its toxicity to become potent venom.

4. Why are insects constrained from growing any larger in their current environment?
   As the size of the insect increases, the ability for their breathing tubes to supply oxygen to all parts of their bodies becomes less and less efficient. Insects must compensate for the increased distances that gases must travel by having wider diameter tubes. But they quickly reach a point where adequate breathing tubes no longer fit within the insect's body (this is especially true for the legs, because they extend the farthest from the insect's body, yet are also very narrow).

Thus there is a maximum size where the tubes, and thus the insect, cannot get any bigger given the current level of oxygen in the atmosphere.

5. Describe the role of opsins and crystallins in the evolution of the vertebrate eye.

Opsins are proteins that allow cells to detect light and transmit this information to other cells. These molecules were present in the ancestor to both cnidarians and bilaterians. Both groups inherited opsins and used them to detect the presence of light and eventually form images. Transparent crystallins evolved, recruited from stable heat-shock proteins, that could focus light. Ultimately, natural selection favored mutations that improved the function of the opsins and the crystallins for vertebrate eyesight.

6. Consider similar traits in both sharks and dolphins. What evidence would you use to identify a trait as being homologous and a different trait as being convergent?

A trait that is homologous between sharks and dolphins is their eyes. It is homologous because it evolved in the common ancestors to both groups, evidence which can be seen in their fossil forms. Further evidence includes the common genes that are involved in the development of the trait as well as the same molecules used for vision in the eye.

One convergent feature is the streamlined shape of the bodies of both sharks and dolphins. This shape was not inherited from a common ancestor. The ancestors of dolphins were tetrapods, and dolphins begin to develop hind limb buds as embryos, just like their legged ancestors. The limb buds stop growing during development, however. Sharks lack the genes necessary to produce hind limbs in the first place. (A second convergence is the evolution of fins and flippers. Fins and flippers evolved independently in dolphins, features their direct terrestrial ancestors did not have.)

7. Are developmental genetic regulatory networks perfectly designed? Explain why or why not.

No, genetic regulatory networks are not designed from scratch; they are pieced together from existing combinations of developmental genes. As a result, few fit the exact needs of any organism perfectly. This can be seen in the deep flaws in complex structures and pathways, such as the blind spot in vertebrate eyes and the incredibly long laryngeal nerve in giraffes.

Test Yourself:
1. b; 2. e; 3. d; 4. d; 5. d; 6. c; 7. b; 8. d; 9. c; 10. d

Chapter 9
SEX AND FAMILY

Check Your Understanding

1. The most important element necessary for natural selection to act is:
   a. Individuals changing to meet their needs.
   b. Variation within a population.
   c. Thousands of years.
   d. All of the above.
   e. None of the above.

2. What is directional selection?
   a. A pattern of natural selection where individuals at one end of the distribution of a trait have greater survival or reproductive success than individuals at the other end.
   b. A pattern of natural selection where individuals near the mean, or center, of the distribution of a trait for a population have greater survival or reproductive success than individuals at the either end.
   c. A pattern of natural selection where individuals near either end of the distribution of a trait for a population have greater survival or reproductive success than individuals near the mean.
   d. All of the above.

3. Which type of evidence provides the most support for natural selection as an important mechanism of evolutionary change?
   a. Observational studies of the changes in a trait over many years.
   b. Experiments that show how selection can cull or favor individuals with certain trait values.
   c. Evaluation of historical natural data that shows changes in a trait over many years and the correlation with environmental variables.
   d. All of the above provide valid evidence for natural selection as an important evolutionary mechanism in the wild.
   e. None of the above provide evidence of natural selection as an important mechanism because they do not actually show evolution occurring.

Learning Objectives for Chapter 9:

- ➢ Identify the genetic consequences of sexual and asexual reproduction.
- ➢ Explain the Red Queen effect.
- ➢ Identify the different investments made in sexual reproduction by males and females.
- ➢ Analyze how the different reproductive strategies of males and females lead to sexual selection.
- ➢ Compare and contrast monogamy with polygyny and polyandry.
- ➢ Analyze how competing interests of males and females in sexual reproduction may influence selection.
- ➢ Explain why organisms might not produce as many offspring as they possibly can.
- ➢ Discuss two trade-offs that might result from investment in reproduction early in life.
- ➢ Examine the relationship between role-reversal and female life-history traits.
- ➢ Explain how conflict between parents might affect life-history evolution within the sexes.
- ➢ Distinguish between the hypotheses that menopause is adaptive and that menopause is a result of life-history trade-offs.

Identify Key Terms

Connect the following terms with their definitions on the right:

anisogamy	A form of sexual selection that arises after mating, when males compete for fertilization of a female's eggs.
gamete	The combining and mixing of chromosomes during the formation of offspring. Two main processes are involved: (1) meiosis halves the number of chromosomes during gamete formation, and (2) fertilization restores the original chromosome number as two gametes fuse to form a zygote.
hermaphrodite	A term that refers to how populations that are coevolving maintain relative fitness, because each population must constantly adapt to the other. Leigh Van Valen borrowed the term from Lewis Carroll's Through the Looking-Glass to refer to biological arms races, such as those between parasites and their hosts.
intersexual selection	Sexual reproduction involving the fusion of two dissimilar gametes; individuals producing the larger gamete (eggs) are defined as female, and individuals producing the smaller gamete (sperm) as male.
intrasexual selection	Competition among the same sex that leads to differential reproductive success.
life history	A mating system where females mate (or attempt to mate) with multiple males.
monogamy	The formation of new individual organisms (offspring).
polyandry	Differential reproductive success resulting from the competition for fertilization.
polygyny	An individual that produces both female and male gametes.
Red Queen effect	A mating system in which one male pairs with one female. Sometimes a male and female form a stable pair bond and cooperate to rear the young, even if either or both partners sneak extra-pair copulations. Rarely does each male mate only with a single female, and vice versa.
reproduction	A cell that fuses with another cell in sexually reproducing organisms during fertilization.
sex	A mating system where males mate (or attempt to mate) with multiple females.
sexual conflict	Competition through attraction to the opposite sex that leads to differential reproductive success.
sexual selection	The investment by an organism into growth and reproduction. Traits include the age at first reproduction, the duration (schedule) of reproduction, the number and size of offspring produced, and lifespan.
sperm competition	The evolution of phenotypic characteristics that confer a fitness benefit to one sex but a fitness cost to the other.

The Tangled Bank Study Guide

Link Concepts

1. Fill in the following concept map with the key terms from the chapter:

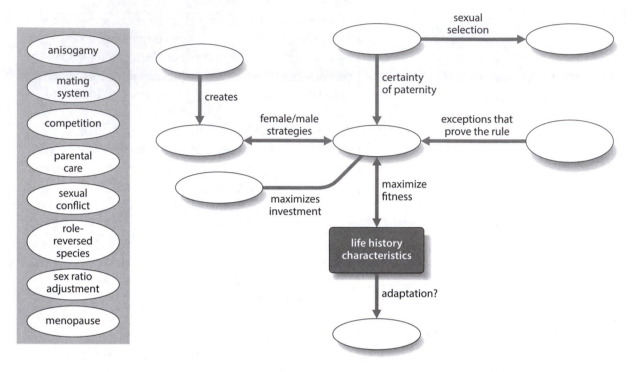

anisogamy

mating system

competition

parental care

sexual conflict

role-reversed species

sex ratio adjustment

menopause

2. Take the above concept map a little deeper. Show how anisogamy can influence the life-history characteristics of males and females differently. Think about concepts like intersexual selection, intrasexual selection, monogamy, polygamy, polygyny, sperm competition, parental care, sexual conflict, and lifetime fitness.

Interpret the Data

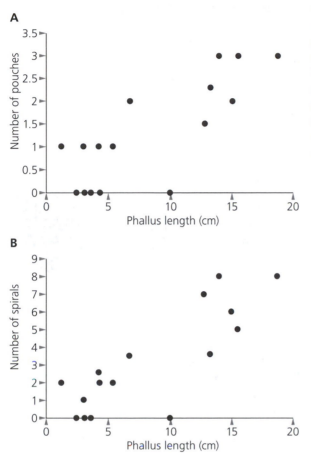

A

B

Patricia Brennan found that among ducks and other wading birds, species where males had long phalluses also had females with many pouches and spirals in their reproductive tracts. (Graph adapted from Brennan et al. 2007.) The mean length of the male's phallus in centimeters for each species is plotted along the x axis, and the mean number of pouches (top) and mean number of spirals (bottom) in the reproductive tracts of females for each species are plotted along the y axes.

What was the longest phallus found for a species?

What was the maximum number of pouches found in females for a species?

What was the maximum number of spirals found in females for a species?

From these graphs, could you predict the number of pouches or spirals you would expect to find in a species where the male phallus length was 9 cm?

Do the relationships between phallus length and number of pouches/spirals shown in the above graphs indicate that male phallus length causes female reproductive tract characteristics?

Games and Exercises

Mating Game

Live from the PBS *Evolution* Library, it's the Mating Game! Test your ability to pick the right mate based on the information the "bachelors" provide. The game is tricky, so read carefully!

http://www.pbs.org/wgbh/evolution/sex/mating/index.html

Mating Game II—Craigslist

Humans are evolving, too, and patterns in our mating preferences, although complex, can be identified. Go to Craigslist.com and examine the personal ads. Although the ads don't represent a true sample of mating preferences in our society, you can still summarize some of the characteristics. For example, how many ads are seeking long-term versus short-term relationships?

Develop a table and tally the number of ads according to gender. Based on your understanding of personal relationships, would you expect age to have an influence? Add age categories to your table. Think about the duration of relationships requested by the people that posted the ads (you probably need to make a new table).

	males	females
18–20		
21–25		
26–30		

- Is duration related to gender?
- Is duration related to age?
- Does gender affect the preference for a mate of a specific age?
- Could age be relevant to female reproductive potential?

Compare ads for heterosexual and homosexual relationships.
- Does the sex of the target audience affect the ads? How?

Develop a hypothesis about sexual selection in humans.

Now, read this:

Ethology **108,** 303—317 (2002)
© 2002 Blackwell Verlag, Berlin
ISSN 0179–1613

Environmental Predictors of Geographic Variation in Human Mating Preferences

Kevin J. McGraw

Department of Neurobiology and Behavior, Cornell University, Ithaca, NY, USA

Abstract

Sexual selection theory classically posits consistent and directional mate-preferences for male traits that provide benefits to females. However, flexible mate-choice tactics may persist within a species when males display multiple desirable features that confer different benefits to females under variable environmental conditions. Ecological factors such as population density, resource demand, and sex ratio can influence the value that female animals place on certain male characteristics across mating environments. In this study, I used human mate-preference data from 'lonely hearts' advertisements in the newspapers of 23 cities in the USA to assess geographic differences in female preferences for male traits (e.g. physical attributes, resource-holding potential, emotional characteristics, personal interests) in relation to these ecological parameters. I found that females placed more emphasis on the resource-accruing ability of prospective mates in densely populated cities and cities having greater resource demands (higher cost of living). In contrast, women from densely populated or resource-demanding cities placed less emphasis on the emotional aspects or personal interests of males. Preferences for physical features were not environmentally linked, but instead were a function of the degree to which females advertised their own physical attractiveness. Collectively, these results suggest that certain mate-choice criteria employed by women are sensitive to variation in local environmental conditions and that variable levels of resource or mate availability may favor different mating tactics across human populations.

Kevin J. McGraw, Department of Neurobiology and Behavior, Cornell University, Ithaca, NY 14853 USA. E-mail: kjm22@cornell.edu

- Does this research affect your hypothesis about age, gender, and duration of relationships in personal ads? Why or why not?

This activity is modified from an exercise developed by Janis Antonovics and Doug Taylor of the Biology Department at the University of Virginia.

http://www.faculty.virginia.edu/evolutionlabs/home.html

Go Online

In this video, the Cornell Lab of Ornithology shares the stunning diversity of plumage and courtship that evolved in one group of birds, the birds of paradise, and explains how these traits would have evolved through natural and sexual selection.

http://ed.ted.com/on/JLk7NshX

The Elaborate Weapons of Beetles—Yes, Beetles

This slideshow from the *New York Times* presents some of the incredible diversity of beetle weapons courtesy of Dr. Douglas Emlen and his colleagues at the University of Montana.

http://www.nytimes.com/slideshow/2009/03/23/science/032409-Armor_index.html

Animal Attraction by Virginia Morell

Bower birds are another amazing example of the lengths males go to to gain access to females. Male bower birds actually build elaborate structures, called bowers, that females inspect. If they find the display suitable, the male may be granted her favor. This *National Geographic* article explores the lengths males will go to, as well as the females' interests, leading to some of the fantastic traits we see in the animal world.

http://science.nationalgeographic.com/science/health-and-human-body/human-body/animal-attraction.html#page=1

Survival of the Sneakiest

The Royal Society of Comics presents this comic strip on how males can cheat and win matings with females. They're not talking about humans, or even big, strong elk, but small (and sexy) crickets.

http://evolution.berkeley.edu/evolibrary/article/0_0_0/sneakermales_01

 Evolution: Why Sex?

Why Sex? is a program in the PBS series Evolution. The program is amazing, and they've highlighted several clips that illustrate the complexities of sexual selection and evolution.

Asexual Reproducers

This segment from *Evolution: Why Sex?* explores this simple question in a species of salamander, *Cnemidophorus tessellatus*, that can reproduce parthenogenetically (eggs can develop without fertilization).

http://www.pbs.org/wgbh/evolution/library/01/5/l_015_01.html

The Red Queen

Another segment of Evolution: Why Sex? examines how sexual reproduction may allow populations to adapt to new selective challenges when asexual populations cannot.

http://www.pbs.org/wgbh/evolution/library/01/5/l_015_03.html

Songbird Infidelity

Many songbirds form what were thought to be monogamous pairs. Scientists soon found out that males were seeking extra pair copulations, but they were surprised to see the extent to which females were slipping out on their male partners. This segment of Evolution: Why Sex? examines how these dalliances influence genetic diversity as well as parental care.

http://www.pbs.org/wgbh/evolution/library/01/6/l_016_07.html

Jacanas and Polyandry

Female wattled jacanas are the competitive sex—not males. Females mate with as many males as possible, and this segment of Evolution: Why Sex? looks at how this role reversal may have evolved.

http://www.pbs.org/wgbh/evolution/library/01/6/l_016_04.html

Creature Courtship by Peter Tyson

PBS *NOVA* explores the diversity of courtship and competition in the animal kingdom, from battling elephant seals to the extravagant duets performed by unrelated male manakins (birds), so that the lead male can mate.

http://www.pbs.org/wgbh/nova/evolution/creature-courtship.html

Sexual Cannibalism

Maydianne Andrade studies Australian redback spiders—a species known for sexual cannibalism. She talks about this intriguing mating behavior in the audio feature from PBS *NOVA*.

http://www.pbs.org/wgbh/nova/nature/profile-maydianne-andrade.html

Common Misconceptions

Evolution Provides Species with the Adaptations They "Need"

Sexually reproducing males and females certainly need to breed if they are to pass on their genes to future generations, but natural and sexual selection do not provide males or females with adaptations simply because they need them. Sexually selected traits arise from processes such as random mutations and the change of gene frequencies from one generation to the next. Sex is the mechanism that permits that transmission.

Evolution Is the Result of Individuals Adapting to Their Environment

Individual organisms undergo change during their own lifetime, and some of these changes are adaptive. But it's generally not possible for these traits to be passed down to successive generations. If you dyed your hair purple, and you found a mate who liked purple hair, you wouldn't expect your children to have purple hair (but the males may have a preference for purple hair if your mate's preference had a genetic basis).

Evolution Has Made Living Things Perfectly Adapted to Their Environment

As impressive as some sexually selected adaptations may be, they are far from perfect. Selection can act only on what already exists, and it operates under tight constraints of physics and development. Bees often "mate" with flowers that only look superficially like female bees, but they still end up pollinating the flower. Humans have very large brains, for example, that have allowed us to become nature's great thinkers, but those big brains also make childbirth much more dangerous for human mothers than for other female primates.

Evolution Happens for the Good of the Species

Just as evolution has no foresight, evolution cannot recognize what is best for a large group of organisms. Sexual selection operates at the level of genes and the reproductive success of individuals.

Contemplate

What would scientists predict about the evolution of life-history traits in a population of lake trout (a popular food fish) after a new lakeside resort is built (some clues can be found in Figure 9.2)?	How might sexual conflict shape the reproductive strategies of male and female gulf pipefish?

Delve Deeper

1. Why is sexual reproduction so widespread when asexual individuals could produce twice as many copies of their genotype as sexual individuals? What would be some costs of reproducing sexually?

2. If two house finch males differ in their expression of a sexually selected ornament, for example the patch of colored feathers on their faces and breasts, why might selection favor females who choose the male on the right, over the others, as a mate?

3. What is polyandry? Describe some of the advantages of polyandry to a female. How could this system of mating increase the chances of survival of her offspring?

4. How does natural selection act to optimize reproductive fitness of individuals in light of life-history trade-offs?

5. Is the following statement an accurate reflection of the current understanding of the evolution of life histories? Why or why not?

"An individual female chooses how many pups to raise because she needs to have high fitness over her lifetime."

Go the Distance: Examine the Primary Literature

Cosima Hotzy and Göran Arnqvist examine an intriguing aspect of sexual selection: when one sex physically injures members of the other sex. Spines on the genitalia of male seed beetles wound female seed beetles during copulation. They test the hypothesis that these spines are adaptive, providing a reproductive edge, so to speak.

- Did Hotzy and Arnqvist find evidence to support the idea that male genital spines are adaptive? Why or why not?

- Did they offer alternative hypotheses?

- Do their conclusions affect scientists' understanding of sexual selection theory? Why or why not?

Hotzy, C., and G. Arnqvist. 2009. Sperm Competition Favors Harmful Males in Seed Beetles. *Current Biology* 19 (5): 404–7. doi:10.1016/j.cub.2009.01.045.
http://www.cell.com/current-biology/abstract/S0960-9822%2809%2900617-4.

Test Yourself

1. Sexual reproduction can speed the spread of adaptations in a population because
   a. Males and females must travel to locate each other; thereby they spread their genes more broadly.
   b. Sexually transmitted diseases reduce fitness of individuals.
   c. Recombination can lead to novel genotypes.
   d. Beneficial mutations can be combined and harmful mutations can be purged.
   e. Both a and b are correct.
   f. Both c and d are correct.

2. The Red Queen effect refers to the fact that
   a. Parasites often kill their hosts and therefore act as potent agents of selection on host populations.
   b. Host immune systems evolve continuously and quickly in an arms race with parasite populations that are also evolving to evade their defenses.
   c. Parasites evolve faster than their hosts.
   d. Social insect colonies often have reproductive queens who actively suppress the reproductive capacity of worker females in the colony.
   e. All of the above.
   f. None of the above.

3. Are unequal gamete sizes relevant for explaining adult behavior?
   a. No. Divergent gamete sizes are a consequence of sexual reproduction, but they are largely irrelevant for understanding adult behavior.
   b. Yes. There typically are insufficient eggs to go around; males end up having to compete for access to them.
   c. Yes. Females sometimes build elaborate nests or burrows in which to place their eggs.
   d. Yes. Both a male and a female gamete are needed to produce viable offspring.
   e. All of the above, except a.
   f. None of the above.

4. Why are traits like the bright colors of a male bird of paradise considered to be honest indicators of male genetic quality?
   a. They stimulate sensory preferences or biases that are intrinsic.
   b. They are very costly for males to produce.
   c. Bright-colored ornaments evolve extremely rapidly and often diverge in form among populations or closely related species.
   d. They accurately reveal unpalatability or distastefulness to predators, so that predators can quickly learn to avoid them.
   e. All of the above. Each of these contributes to the "honesty" of sexually selected male ornaments.
   f. None of the above.

5. Which statement is the best way to explain why an understanding of life-history trade-offs is important when examining the fitness of an organism?

   a. Because individuals need to live longer, so they will not produce the maximum number of offspring each season.
   b. Because when individuals need offspring, they can choose whether to reproduce or not.
   c. Because individuals may not be producing the maximum number of offspring they possibly can in any particular breeding season.
   d. Because in any particular breeding season, individuals may need to produce fewer offspring to enhance their own survival.

6. What was Austad's prediction regarding opossums on Sapelo Island?
   a. Opossums on Sapelo Island would mature later and have fewer offspring per season than opossums on the mainland.
   b. Opossums on Sapelo Island would have more offspring per season than opossums on the mainland because there were no predators.
   c. Natural selection would favor opossums with more stretchable muscle fibers on Sapelo Island.
   d. Opossums on Sapelo Island would have higher fitness than opossums on the mainland.
   e. Opossums on Sapelo Island would have lower fitness than opossums on the mainland.

7. Do natural selection and sexual selection always act similarly to produce the highest fitness for both males and females?
   a. No. Natural selection acts on females to produce the maximum number of offspring, and sexual selection acts on males to help them outcompete rivals.
   b. No. Natural selection and sexual selection can have competing effects if the competition for mates is extreme.
   c. Yes. Because males and females both benefit by producing the maximum number of offspring, natural selection and sexual selection act to produce the maximum number of offspring for males and females over their lifetimes.
   d. Yes. Males and females both need to produce the maximum number of offspring because natural selection and sexual selection weed out those individual males and females that cannot.

8. Why do male gobies dig up and devour eggs in their nests?
   a. Because the male needs the nutrition, and he can always mate again later.
   b. Because if there is not enough oxygen in the water, the young won't survive anyway.
   c. Because male gobies that devour some of their eggs in response to poor environmental conditions tend to have more surviving offspring than male gobies that do not.
   d. The male eats the young for the betterment of the species.
   e. The male eats the young when he is unsure of paternity.

9. Why would some scientists argue that menopause simply results from life-history trade-offs?
   a. Because everything is a trade-off, and it makes sense that a female's inability to breed late in life must be a trade-off.
   b. Because natural selection cannot act on individuals that no longer reproduce.
   c. Because genetic imprinting by males can cause females to become infertile.
   d. Because the mechanism that leads to infertility in menopausal women, i.e., damage to a female's eggs over time, is the same in other species.
   e. All of the above.

The Tangled Bank Study Guide

Check Your Understanding

1. The most important element necessary for natural selection to act is:

   a. Individuals that can change to meet their needs.

   Incorrect. Although individuals can change to meet their needs, these changes have no genetic basis—evolution cannot identify an individual's needs. Natural selection is a population-based phenomenon that requires heritable variation among individuals that confers some advantage or disadvantage. Within a population, those that do better, for example, are more likely to reproduce and have offspring that share some of the variable traits that led to that success. Over time, individuals with those successful traits become more common than those without— the essence of natural selection (see Chapter 6).

   b. Variation within a population.

   Correct. Heritable variation among individuals serves as a vital raw material for natural selection. Natural selection is a population-based phenomenon that requires heritable variation among individuals that confers some advantage or disadvantage. Within a population, those that do better, for example, are more likely to reproduce and have offspring that share some of the variable traits that led to that success. Over time, individuals with those successful traits become more common than those without—the essence of natural selection (see Chapter 6).

   c. Thousands of years.

   Incorrect. Although lineages can evolve in relatively short time periods, evolution often requires a significantly longer period than thousands of years. For example, fast generation times (such as in insects, viruses, and bacteria) or when natural selection is particularly strong (such as selective sweeps, see Chapter 6) can result in rapid evolution, but the evolution of hominids took millions of years (see Chapter 4).

   d. All of the above.

   Incorrect. Individuals can change to meet their needs, but these changes have no genetic basis—evolution cannot identify an individual's needs (Chapter 2). Also, evolution usually takes time. Although lineages can evolve in relatively short time periods, evolution often requires a significantly longer period than thousands of years. Variation within a population is essential for natural selection and can be considered the most important element for it to act (Chapter 6).

   e. None of the above.

   Incorrect. It is correct to say that individuals changing to meet their needs is not important to natural selection, and that thousands of years is not the most important element. However, variation within a population is essential for natural selection and can be considered the most important element for it to act (Chapter 6).

2. What is directional selection?

   a. A pattern of natural selection where individuals at one end of the distribution of a trait have greater survival or reproductive success than individuals at the other end.

   Correct. When natural selection favors individuals with a phenotypic trait—large body size of fishes in a pond, for example—large individuals do relatively better than small individuals. Over time, the large individuals have more surviving offspring than the small, and the proportion of large individuals in the population increases. Graphically, the distribution of the sizes of fish in the population would gradually shift toward larger and larger individuals.

   b. A pattern of natural selection where individuals near the mean, or center, of the distribution of a trait for a population have greater survival or reproductive success than individuals at the either end.

   Incorrect. When natural selection favors individuals with a phenotypic trait—moderate body size of fishes in a pond, for example— large individuals and small individuals do relatively worse than individuals close to the mean. This selective advantage can result because different selective agents focus on large and small individuals, while moderate individuals fare better overall. Graphically, the distribution of the sizes of fish in the population would gradually squeeze in toward the center of the distribution. Directional selection occurs when individuals with a phenotypic trait at one end of the distribution do relatively better than individuals at the other end.

  c. A pattern of natural selection where individuals near either end of the distribution of a trait for a population have greater survival or reproductive success than individuals near the mean.

Incorrect. When natural selection favors individuals with extreme phenotypic traits—both small and large body size of fishes in a pond, for example—individuals whose traits are near the mean do relatively worse than individuals whose traits are closer to one end or the other. This selective advantage can result because different selective agents favor large and small individuals—perhaps for different reasons—while moderate individuals fare less well. Graphically, the distribution of the sizes of fish in the population would gradually push toward both tails of the distribution, away from the center. Directional selection occurs when individuals with a phenotypic trait at one end of the distribution do relatively better than individuals at the other end.

  d. All of the above.

Incorrect. Directional selection occurs when natural selection favors individuals with a phenotypic trait at one end of the distribution of a trait. Stabilizing selection can narrow the range of body sizes by favoring individuals at the center of the distribution, whereas disruptive selection selects against the population mean, favoring individuals at either end of the distribution.

3. Which type of evidence provides the most support for natural selection as an important mechanism of evolutionary change?

  a. Observational studies of the changes in a trait over many years.

Correct, but so are other answers. Observational studies can provide important clues as to the characters that are changing and have been vital in the study of evolution, especially when combined with other ways of addressing the mechanism of natural selection (as opposed to genetic drift; Chapter 6).

  b. Experiments that show how selection can cull or favor individuals with certain trait values.

Correct, but so are other answers. Experiments that cull or favor individuals can

directly address the role of natural selection as a mechanism for evolutionary change, but some organisms (*E. coli*) are easier to experiment with than others (oldfield mice). Scientists can also come up with creative experiments that allow them to test how natural selection is operating within populations of organisms where they can't morally cull or favor individuals, like humans (Chapter 6).

  c. Evaluation of historical natural data that shows changes in a trait over many years and the correlation with environmental variables.

Correct, but so are other answers. Historical natural data can offer a treasure trove of evidence. When scientists can find strong correlations with environmental variables, especially when they examine the correlations in different contexts, or relationships that are repeated in many different populations, they can infer mechanisms such as natural selection. This evidence can also be compared with modern studies adding to the evidence for the important role natural selection plays in evolutionary change (Chapter 6).

  d. All of the above provide valid evidence for natural selection as an important evolutionary mechanism in the wild.

Correct. Scientists have accumulated so much evidence from observational studies, experiments, and historical data that they no longer question the importance of natural selection as a mechanism of evolutionary change. All this evidence has generated exciting new questions, however, such as how phenotypes and genotypes interact in the face of natural selection (Chapter 7), how the drive to reproduce interacts with natural selection (this chapter), and how other species (Chapter 12) and large-scale geologic events influence the outcomes of natural selection (Chapter 11).

  e. None of the above provide evidence of natural selection as an important mechanism because they do not actually show evolution occurring.

Incorrect. Many observational and experimental studies have actually shown evolutionary change within populations. Organisms, such as *E. coli*, *Drosophila* (fruit flies), and even house mice, are regularly used in laboratory experiments to show how different agents of natural selection can affect

The Tangled Bank Study Guide

the evolution of character traits. Scientists can even see evolution occurring in the field, such as resistance to pesticides evolving in mosquitoes in France, and salmonid fish changing to become sexually mature at an earlier age as a result of overfishing (see Chapter 6).

Identify Key Terms

anisogamy	Sexual reproduction involving the fusion of two dissimilar gametes; individuals producing the larger gamete (eggs) are defined as female, and individuals producing the smaller gamete (sperm) as male.
gamete	A cell that fuses with another cell in sexually reproducing organisms during fertilization.
hermaphrodite	An individual that produces both female and male gametes.
intersexual selection	Competition through attraction to the opposite sex that leads to differential reproductive success.
intrasexual selection	Competition among the same sex that leads to differential reproductive success.
life history	The investment by an organism into growth and reproduction. Traits include the age at first reproduction, the duration (schedule) of reproduction, the number and size of offspring produced, and lifespan.
monogamy	A mating system in which one male pairs with one female. Sometimes a male and female form a stable pair bond and cooperate to rear the young, even if either or both partners sneak extra-pair copulations. Rarely does each male mate only with a single female, and vice versa.
polyandry	A mating system where females mate (or attempt to mate) with multiple males.
polygyny	A mating system where males mate (or attempt to mate) with multiple females.
Red Queen effect	A term that refers to how populations that are coevolving maintain relative fitness, because each population must constantly adapt to the other. Leigh Van Valen borrowed the term from Lewis Carroll's Through the Looking-Glass to refer to biological arms races, such as those between parasites and their hosts.
reproduction	The formation of new individual organisms (offspring).
sex	The combining and mixing of chromosomes during the formation of offspring. Two main processes are involved: (1) meiosis halves the number of chromosomes during gamete formation, and (2) fertilization restores the original chromosome number as two gametes fuse to form a zygote.
sexual conflict	The evolution of phenotypic characteristics that confer a fitness benefit to one sex but a fitness cost to the other.
sexual selection	Differential reproductive success resulting from the competition for fertilization.
sperm competition	A form of sexual selection that arises after mating, when males compete for fertilization of a female's eggs.

Link Concepts

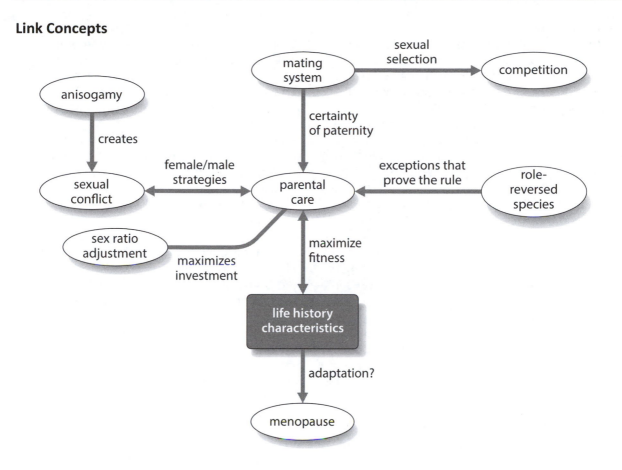

Interpret the Data

- What was the longest phallus found for a species?
  Approximately 19 cm.
- What was the maximum number of pouches found in females for a species?
  3 pouches.
- What was the maximum number of spirals found in females for a species?
  8 spirals.
- From these graphs, could you predict the number of pouches or spirals you would expect to find in a species where the male phallus length was 9 cm?
  Yes, you could predict that in a species where the male had a mean phallus length of 9 cm, the female reproductive tract would likely have between 0 and 2 pouches, and 0 to 4 spirals.
- Do the relationships between phallus length and number of pouches/spirals shown in the above graphs indicate that male phallus length causes female reproductive tract characteristics?
  No. The relationships are correlations; they show a positive relationship between the variables (male phallus length and either number of pouches or number of spirals in the female reproductive tract) but not that male phallus length causes either one of the female characteristics. Selection is clearly acting on both males and females, but causative relationships would require a mechanism for how the male phallus length actually made the female reproductive tract change.

Delve Deeper

1. Why is sexual reproduction so widespread when asexual individuals could produce twice as many copies of their genotype as sexual individuals? What would be some costs of reproducing sexually?

   Sexual reproduction is so common because of the reproductive advantage it offers. Sexual reproduction allows the combination of genomes, bringing together beneficial adaptations much faster than if they arose separately. Recombination during meiosis also creates new genotypes with various combinations of alleles, and recombination can lead to the exclusion of deleterious mutations from some genotypes. The genetic diversity created by sexual reproduction can lead to rapid evolution because selection has so much variation to act upon.

   Sex can be costly because it facilitates the transmission of diseases and parasites through sexual contact. Also, finding and assessing a mate can cost time that could be spent foraging and expose individuals to a greater number of predators.

2. If three house finch males differ in their expression of a sexually selected ornament, for example the patch of red feathers on their faces and breasts, why might selection favor females who choose the male on the right, over the others, as a mate?

   The female would benefit if the brightest male also was the best quality male. Bright red feathers may be a reliable indicator of good genes if developing red feathers comes at some kind of cost that can't be faked (good genes hypothesis). For example, the pigments may come from a food source that is particularly difficult to acquire because males have to compete with each other for access. Or, if bright colors attract predators, then only the most fit males may be able to survive despite the handicap of being bright. A bright red male may be unusually high quality (handicap hypothesis). Another reason selection may favor females who choose specific males may be that females benefit by picking bright males as mates because these males are likely to sire the brightest male offspring. Her sons would therefore be especially likely to mate themselves (sexy son hypothesis).]

3. What is polyandry? Describe some of the advantages of polyandry to a female. How could this system of mating increase the chances of survival of her offspring?

   Polyandry occurs when females mate with more than a single male. By mating with multiple males, a female increases her chances of producing offspring that combine her genes with the highest quality male genes (more males equals more possible combinations of alleles, some of which will be more beneficial than others).

4. How does natural selection act to optimize reproductive fitness of individuals in light of life-history trade-offs?

   Individuals should be maximizing their reproductive effort, but investment in reproduction often involves trade-offs between reproduction and growth/body maintenance, or trade-offs between the ability to breed early or later in life. Although individuals experience the trade-offs associated with the schedule and duration of key events in their lives, those trade-offs are only subject to natural selection if (1) there is variability among individuals, (2) that variability is heritable, and (3) that variability confers some advantage or disadvantage to the individuals in terms of reproduction or survival. Some mutations may affect an individual's sensitivity to external conditions, making reproduction more or less likely. Others may be more or less sensitive to hormones that influence behaviors, such as the release of ova in females. Still other mutations may lead to higher or lower levels of egg dumping, for example, or egg consumption. Selection should favor the optimal trade-off that maximizes the number of offspring that survive to maturity over the course of an organism's entire life.

5. Is the following statement an accurate reflection of the current understanding of the evolution of life histories? Why or why not? "An individual female chooses how many pups to raise because she needs to have high fitness over her lifetime."

   No, the statement does not accurately reflect the current understanding of evolutionary biologists. Selection would favor the optimal trade-off that maximized reproductive fitness but not because of an individual's need.

Although the individual female is making a "choice," that choice is not governed by a sense of need or an understanding of future breeding opportunities. Different species, even different populations, will respond differently to the selection pressures that shape their environment. Much like with behavioral plasticity, natural selection can shape mechanisms that affect how individuals allocate resources to reproduction or growth/survival in response to conditions. Heritable mechanisms that affect the lifetime fitness of individuals will either be more or less represented in future generations based on whether they were successful or not.

Test Yourself:
1. f; 2. b; 3. b; 4. b; 5. c; 6. a; 7. b; 8, c; 9. d

Chapter 10
DARWIN'S FIRST QUESTION: THE ORIGIN OF SPECIES

Check Your Understanding

1. How does sexual selection differ from natural selection?
   a. Sexual selection only acts on males, whereas natural selection acts on populations.
   b. Variation among the choosy sex is not necessary for sexual selection to operate; variation among all individuals is necessary for natural selection to operate.
   c. Sexual selection does not really differ from natural selection; both require heritable variation in traits that confer some fitness differences to individuals possessing those traits.
   d. Sexual selection does not really differ from natural selection, but the optimum trait value under sexual selection can often be at odds with the optimum trait value under natural selection.
   e. Both c and d.

2. What are clades?
   a. Groups made up of organisms and all of their descendants.
   b. Hierarchies nested according to shared traits that are inherited from recent common ancestors.
   c. Groups of organisms that comprise a taxonomic unit.
   d. Groups of living organisms that share a phenotypic trait or character state.
   e. Both a and b.

3. How accurate is the current estimate for the age of the Earth (4.567 billion years)?
   a. Not accurate at all because it is just an estimate.
   b. Only slightly accurate because radiometric dating requires knowing decay rates, and no one has measured decay rates directly.
   c. Somewhat accurate, but different dating techniques give different results.
   d. Fairly accurate because scientists have been replicating the experiments and reevaluating the evidence for decades.

Learning Objectives for Chapter 10:

➤ Compare and contrast the phylogenetic species concepts and the biological species concepts.
➤ Explain how isolating barriers contribute to speciation.
➤ Differentiate between allopatric and sympatric speciation.
➤ Compare and contrast geographic isolation with speciation.
➤ Explain how ring species can lead to speciation in the absence of allopatry.
➤ Explain why the rate of speciation varies among organisms.
➤ Describe an example of a cryptic species.
➤ Compare species concepts applied to eukaryotes and bacteria.

Identify Key Terms

Connect the following terms with their definitions on the right:

allopatry	Intrinsic properties of organisms that reduce the likelihood of interbreeding between individuals of different populations.
allopolyploidy	Groups of organisms that are genetically distinct and do not interbreed but are morphologically almost indistinguishable.
biological species concept	Populations occurring in separate, nonoverlapping geographic areas (i.e., they are separated by geographic barriers to gene flow).
cryptic species	Species are the smallest possible groups whose members are descended from a common ancestor and who all possess defining or derived characteristics that distinguish them from other such groups.
geographic barriers	Populations occurring in the same geographic area.
horizontal gene transfer	The evolutionary process by which new species arise, resulting in one evolutionary lineage that splits into two or more lineages.
isolating barrier	Features of the environment that physically separate populations from each other.
phylogenetic species concept	An aspect of the environment, genetics, behavior, physiology, or ecology of a species that reduces or impedes gene flow from individuals of other species.
reproductive barriers	Species are groups of actually (or potentially) interbreeding natural populations that are reproductively isolated from other such groups.
reproductive isolation	The occurrence of more than two paired chromosomes as a result of interspecific hybridization.
ring species	A connected series of populations, each of which can interbreed with its neighboring populations, that has diverged sufficiently across its range that the populations at the ends of the series are too different to interbreed.
speciation	Any process in which genetic material is transferred to another organism without descent.
sympatry	Occurs when reproductive barriers prevent or strongly limit reproduction between populations. The result is that few or no genes are exchanged between the populations.

Link Concepts

1. Fill in the following concept map with the key terms from the chapter:

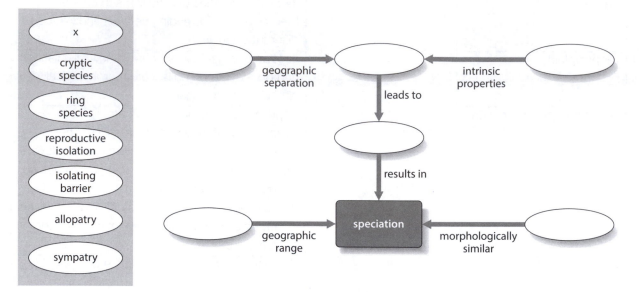

2. Develop your own concept map that shows the relationship between species concepts and speciation. Think about key differences among species concepts and how different taxa may or may not fit those concepts. Then think about when those different concepts can be applied and how the outcomes might affect understanding speciation.

Interpret the Data

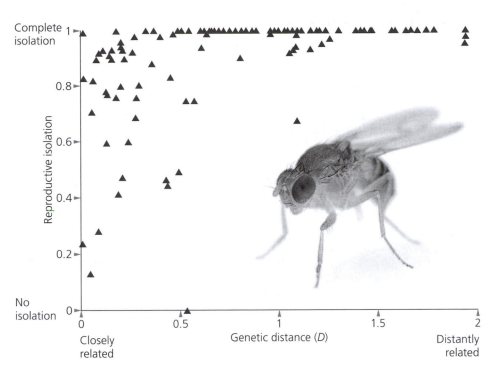

This graph shows how reproductive isolation evolved among species of *Drosophila*. The genetic distance (D) between two species increases with time. (Adapted from Coyne and Orr 2004.)

Were *Drosophila* species pairs more likely to hybridize when they had a genetic distance < 0.5 or > 0.5?

Did *Drosophila* species pairs hybridize even if they weren't closely related?

If it takes roughly a million years for D to reach a value of 1, when would you consider *Drosophila* species to be reproductively isolated?

Games and Exercises

The**Cornell**Lab of Ornithology

Singing a Song of Speciation

A number of species concepts have been proposed, and no one definition likely fits all species. For example, the biological species concept considers species as groups of actually (or potentially) interbreeding natural populations that are reproductively isolated from other such groups. The phylogenetic species concept considers species as the smallest possible groups whose members are descended from a common ancestor and who all possess defining or derived characteristics that distinguish them from other such groups. Focusing on the correct definition isn't as interesting as exploring the processes by which species evolve, the forces affecting populations, and the cohesion of these evolutionary units.

The Cornell Lab of Ornithology developed this multimedia activity to illustrate some of the difficulties with our concepts of species (http://orb.birds.cornell.edu/orb/wp-content/uploads/2013/05/InstructorGuide-SpeciesConcepts.pdf). Eastern and western meadowlarks and blue-winged and golden-winged warblers are all recognized as distinct species. Examine the video and songs of these four species.

The first pair of species is the meadowlarks:
http://macaulaylibrary.org/video/415063
http://macaulaylibrary.org/video/435410

The second pair of species is the warblers:
http://macaulaylibrary.org/video/436872
http://macaulaylibrary.org/video/435383

- How different are the species in each pair?
- Do they look alike?
- Would you predict that the two meadowlark species might hybridize?
- What about the warblers?

Download the Raven Viewer (http://macaulaylibrary.org/raven-viewer) to analyze the bird songs using visual representations of the audio. Once the viewer is downloaded, use the visualize icon (it looks like a mini-waveform ⊞) to examine each species' song.

- Can you see any quantifiable differences in the birds' songs?

- Which species pair has greater differences, meadowlarks or warblers?

The lab has a number of songs of individuals of each species available.
- What kind of variation among individuals can you see?

Now examine the species accounts for eastern and western meadowlarks and blue-winged and golden-winged warblers at AllAboutBirds.org.

http://www.allaboutbirds.org/guide/eastern_meadowlark/id

 http://www.allaboutbirds.org/guide/Western_Meadowlark/id

http://www.allaboutbirds.org/guide/Blue-winged_Warbler/id

http://www.allaboutbirds.org/guide/Golden-winged_Warbler/id

- How much geographic overlap is there between eastern and western meadowlarks?

- What about blue-winged and golden-winged warblers?

What do the species accounts say about hybridization? Are you surprised given your predictions about how similar or different the species pairs look? What about the differences in their songs? What can you conclude about our understanding of species concepts based on this activity?

Ring Species

Ring species are incredibly powerful evidence of the variation that can lead to speciation over time. See if you can explain ring species with this semantic map. Semantic mapping is simply a strategy for representing concepts in graphic form. They are one way of examining the relationships that comprise a concept like ring species. Try developing similar semantic maps for allopatric and sympatric speciation.

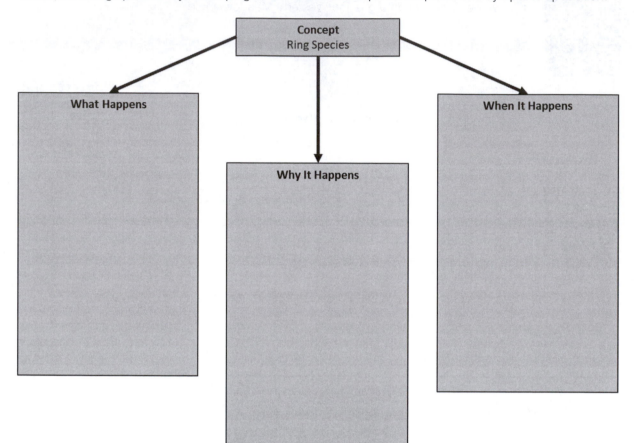

For more information about ring species and speciation, see Darren Irwin's actionbioscience page at http://www.actionbioscience.org/evolution/irwin.html.

ENSI also has a great activity to help you see how biogeography can inform speciation using different subspecies of the California Salamander. You can examine the different traits of populations and map where they've been found on a topographic map of California. It's a great visual exercise!

http://www.indiana.edu/~ensiweb/lessons/step.sp.html

Go Online

The Speciation Song

Cheryl Van Buskirk has written and animated a song about speciation. Watch the Speciation Song here:
http://www.youtube.com/watch?v=WDPsZPKSEFg

She provides a little more explanation of the process here:
http://www.youtube.com/watch?v=PKb8Yi5xzhE

Connecting Concepts: Interactive Lessons in Biology

The University of Wisconsin-Madison produced this animated module on species and speciation. You can test your understanding of species and speciation, then apply that understanding to two real-world examples.

http://ats.doit.wisc.edu/biology/ev/sp/sp.htm

Speciation: Of Ligers and Men—Crash Course Biology #15

Hank Green explains speciation in this crash course in biology. He discusses the kinds of barriers that can arise that lead to reproductive isolation—prezygotic and postzygotic, allopatric and sympatric speciation, and hybridization.

http://www.youtube.com/watch?v=2oKlKmrbLoU

Dr. William Hanna developed some great animations about speciation for his course Biological Principles II at Massasoit Community College.

Allopatric Speciation

This animation illustrates vicariance and dispersal and discusses when speciation doesn't occur, and when it does.

http://faculty.massasoit.mass.edu/whanna/122/page4/page28/page50/page50.html

Sympatric Speciation

This animation offers a great explanation of meiosis and ploidy, and specifically how speciation by allopolyploidy can occur.

http://faculty.massasoit.mass.edu/whanna/122/page4/page28/page49/page49.html

Hummingbird Species in the Transitional Zones

PBS *Evolution* "*Darwin's Dangerous Ide*a" explores the hummingbirds and their habitats on the east slope of the Andes Mountains in Ecuador. Biologists studying these amazing birds are discovering that ecological differences can drive speciation. Small changes in bill length can lead to big changes in evolutionary lineages.

http://www.teachersdomain.org/resource/tdc02.sci.life.evo.hummingbird/

Salamander Ring Species

PBS *NOVA*, "*Evolution in Action*" offers a great video about Ensatina salamanders in a classic example of ring species:

http://video.pbs.org/video/1300397304/

 Speciation in Real Time

Explore recent research on speciation in the central European blackcap. This website includes a video from National Evolutionary Synthesis Center (NESCent).

http://evolution.berkeley.edu/evolibrary/news/100201_speciation

Speciation: An Illustrated Introduction

In this animated video, the Cornell Lab uses the biological species concept to explain the diversity of birds, and species in general. The video includes stunning images of the birds of paradise.

http://ed.ted.com/on/R4PiqPn1

New Tools for Separating Geographical and Ecological Isolation

Hot off the press—scientists have developed new statistical techniques for examining genetic differentiation due to geographical isolation versus ecological isolation.

http://onlinelibrary.wiley.com/doi/10.1111/evo.12193/abstract

Common Misconceptions

Evolution Is Not Entirely Random

Chance is certainly an important component of evolution. Mutations—harmful, beneficial, and even neutral—are random, and within an individual they arise because of chance. But whether a particular mutation spreads throughout a population is not solely due to chance, especially if that mutation affects the reproductive success of the individuals that have it. Natural selection is definitely not random; it does act on the raw heritable variation that arises by chance. In fact, the nonrandomness of evolution by natural selection is amazingly apparent. Whales and fish occupy similar ecological niches in the same physical environment space. They face similar selection pressures, and

the result has been convergence on similar shapes. The natural world is full of evidence for the nonrandom nature of evolution.

Evolution Can and Has Been Observed

Evolutionary biologists have been observing evolution for centuries, in artificial breeding, in the lab, and even in the wild. Artificial breeding may not seem like evolution, but the process has definitely changed the frequency of alleles over time. The result has been an amazing diversity, and over time these groups may or may not be able to interbreed. In the lab, scientists have used model organisms, like *Drosophila*, to show sympatric speciation, and even to show when reproductive isolation is likely to occur. In fact, in the lab scientists have been able to show many mechanisms that can cause gene frequencies to change over time, including predation and sexual selection. In the wild, evolutionary biologists have observed the evolution of mouse coat color, allopolyploidy in plants, and the evolution of resistance to pesticides and herbicides, for example.

Observing evolution isn't limited to just what we see before our eyes. Observation in evolution, and science in general, uses indirect evidence as well, and that indirect evidence is overwhelming. More

importantly, evolutionary theory makes predictions about what we expect to see, and scientists can test those predictions with both direct and indirect evidence. We can predict where we would expect to find common ancestors in geologic formations; we can predict when life-history traits should differ among populations; and we can predict how new pathogens that affect human health may evolve.

Evolution Is NOT a March of Progress

People often ask, if humans evolved from apes, then why are there still apes? No evolutionary biologist would ever claim that humans evolved from modern apes—humans and apes share a common ancestor. The lineage of that common ancestor gave rise to both human and ape lineages, and that is why evolutionary biologists consider apes our closest living relatives.

People often identify the "March of Progress" illustration as a true depiction of the relationship between modern apes (chimps) and humans, but evolution rarely occurs in a straight line. Phylogenetic trees are much better tools for visualizing how lineages split and diverge.

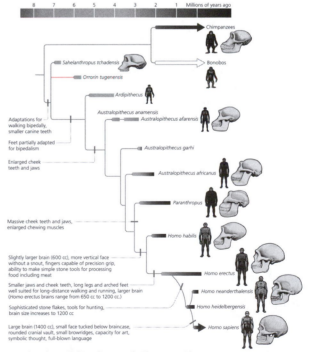

And nothing in evolutionary theory suggests that when one group splits from another, the first group has to go extinct to make way for the new group. Think about allopatric speciation—the rise of a geographic barrier, such as a river changing course, can lead to speciation in the group of individuals isolated by the new course. That doesn't mean the original group has to go extinct for the two groups to coexist in time (or even space if the geographic barrier is removed in the future).

So the common ancestor of apes and humans may have gone extinct, but apes and humans (as well as all related species) can coexist because lineages split and change. Some lineages go extinct, some do not, and evolutionary biologists use the clues left behind from this great diversity as evidence to reconstruct our complex past.

Transitional Fossils Are Everywhere

Scientists expect gaps in the fossil record because the fossil record preserves only a tiny fraction of the history of life, and discovering fossils of new species is a slow, arduous process. Since the publication of the *On the Origin of Species*, however, thousands and thousands of new fossils have been discovered, including many fossils of human ancestors.

So what counts as a transitional fossil? To evolutionary biologists, it's a fossil that has some characteristics of one lineage, and some characteristics of another, and perhaps some characteristics of its own lineage. It is an intermediate—a position that depends on the categories being compared. So *Archaeopteryx* is a transitional fossil between dinosaurs and birds if, like scientists, you understand that dinosaurs and birds are very loose categories. No paleontologist expects to find a fossil that is directly transitional between *Tyrannosaurus rex* and *Cardinalis cardinalis* (the northern cardinal). But, scientists have started looking for fossils with characteristics the two groups share—like feathers—and they have found a ton of new fossils (literally) that have features that make them intermediate—or transitional—to birds and dinosaurs.

Likewise, the new fossil hominids that are being discovered regularly function as transitional lineages between apes and humans.

Contemplate

What does the plethora of cryptic species being discovered using genetic tools say about our human understanding of what a species is?	Given your experience with natural populations, would you suggest that allopatric or sympatric speciation is more common?

Delve Deeper

1. Do geographic isolating barriers function equally under the phylogenetic and biological species concepts?

2. Why would different kinds of bacteria and archaea be difficult to identify as different species using the biological species concept?

3. Explain how firefly flashes can act as reproductive barriers.

4. Predict what would happen if you took a female greenish warbler from the oldest, extreme southern end of its distribution and transplanted her to the most northern end of the warbler's range in Siberia, where the two extremes of the ring species overlap. Would she be able to successfully mate with either an east or west Siberian warbler? Both? Neither? Explain why referring to both behavior and genetics.

5. What are the similarities and differences between allopatric and sympatric speciation?

6. Under what ecological and evolutionary conditions is sympatric speciation most likely?

7. Which species concept would you apply to plants that hybridize relatively frequently, such as salsify flowers (*Tragopogon*), often resulting in allopolyploidy?

Go the Distance: Examine the Primary Literature

The neotropical skipper butterfly, *Astraptes fulgerator*, was first described in 1775, and it was considered a common species that was variable and wide ranging. In 2004, Paul Hebert and his colleagues examined Astraptes fulgerator to understand biodiversity in this biologically important part of the world, and they found that Astraptes fulgerator was likely more than a single, generalist species—it was 10 distinct species.

- What evidence did Hebert and his colleagues use to evaluate the differences among skipper species?

- Was only a single type of evidence used to develop their phylogenetic tree?

- Why do Hebert and his colleagues suggest that adults are so similar in coloration?

Hebert, P. D. N., E. H. Penton, J. M. Burns, D. H. Janzen, and W. Hallwachs. 2004. Ten Species in One: DNA Barcoding Reveals Cryptic Species in the Neotropical Skipper Butterfly *Astraptes fulgerator*. *Proceedings of the National Academy of Sciences* 101 (41): 14812–17. doi:10.1073/pnas.0406166101. http://www.pnas.org/content/101/41/14812.

Test Yourself

1. In which case would the biological species concept not be useful?
   a. Different kinds of birds occurring sympatrically with very different appearances.
   b. A group of lizards reproducing asexually.
   c. Big cats living in Asia and big cats living in Africa.
   d. Both a and b.
   e. Both b and c.

2. Why is defining the concept of species such a difficult task?
   a. Species are constantly evolving.
   b. Species are often defined in relation to research methods.
   c. Species are fixed taxonomic units—the difficulty arises from asexually reproducing organisms.
   d. Both a and b.
   e. Both b and c.

3. According to Figure 10.4, how are Monostrea corals reproductively isolated?
   a. They live in different habitats.
   b. One species spawns in the morning, the other in the evening.
   c. The distribution of spawning times is completely nonoverlapping.
   d. Species release their gametes for different lengths of time.
   e. The gametes are dispersed by ocean currents at different rates.

4. Are reproductive isolating barriers more important in allopatric speciation or sympatric speciation?
   a. Allopatric speciation.
   b. Sympatric speciation.
   c. Reproductive isolating barriers are important in both allopatric and sympatric speciation.
   d. Only geographic isolating barriers are important in speciation.

5. Which is the most likely order of events that could lead to allopatric speciation?
   a. Geographic separation, then genetic divergence, then reproductive isolation.
   b. Genetic divergence, then geographic separation, then reproductive isolation.
   c. Genetic divergence, then reproductive isolation, then geographic separation.
   d. Geographic separation, then reproductive isolation, then genetic divergence.

6. Which of these statements is TRUE about ring species?
   a. Individuals can easily migrate across the center of the ring.
   b. Alleles can originate at one end and flow all the way to the other end.
   c. Any population in the ring can successfully reproduce with any other population in the ring.
   d. Individuals can migrate anywhere within the geographic range of the ring but do not leave that area.
   e. Ring species include populations completely isolated from allelic exchange.

7. Which of these statements about allopolyploidy is FALSE?
   a. Allopolyploidy occurs only in plants.
   b. Allopolyploidy is the doubling of chromosomes as a result of hybridization.
   c. Allopolyploidy can quickly lead to speciation.

d. In a phylogeny of the Tragopogon, an allopolyploidy event would be represented as a node.

8. Which of these statements about the speed of speciation is FALSE?
   a. Speciation can take millions of years.
   b. Speciation can happen in a single generation.
   c. Speciation in plants can result because of hybridization.
   d. Speciation often happens faster in flowering plants than in animals.
   e. None of the above is false.

9. What are cryptic species?
   a. Metapopulations of organisms that exchange alleles frequently enough that they comprise the same evolutionary lineage but are almost indistinguishable morphologically.
   b. Groups of organisms that are genetically distinct and do not interbreed, but are almost indistinguishable morphologically.
   c. The smallest possible groups whose members are descended from a common ancestor and who all possess defining or derived characteristics that distinguish them from other such groups, even though they are almost indistinguishable morphologically to humans.
   d. Groups of organisms that have converged on similar adaptations so much so that they are almost indistinguishable.

10. Why is horizontal gene transfer an important factor in defining microbial species?
   a. Because horizontal gene transfer, along with rapid diversification, undermines any universal species concept.
   b. Because homologous recombination could be considered microbial "sex."
   c. Because horizontal gene transfer transforms microbial genomes into mosaics difficult to classify.
   d. All of the above.
   e. None of the above.

Answers for Chapter 10
Check Your Understanding

1. How does sexual selection differ from natural selection?

   a. Sexual selection only acts on males, whereas natural selection acts on populations.

   Incorrect. Natural selection and sexual selection both act on the heritable variation among individuals within a population; sexual selection can be considered a subset of natural selection. It's the differences in fitness that arise from competition over access to reproduction. Although sexual selection typically involves males competing with rival males for access to females, occasionally it can work the other way around, with females competing for access to males (see Chapter 9).

   b. Variation among the choosy sex is not necessary for sexual selection to operate; variation among all individuals is necessary for natural selection to operate.

   Incorrect. Natural selection and sexual selection both act on the heritable variation among individuals within a population; the difference is that sexual selection is the differential reproductive success that results because individuals vary in traits that affect their ability to compete for fertilizations. Sexual selection affects the sex with the greater variance in reproductive success, but both the trait and the preference for that trait can vary (see Chapter 9).

   c. Sexual selection does not really differ from natural selection; both require heritable variation in traits that confer some fitness differences to individuals possessing those traits.

   Correct, but so are other answers. Natural selection and sexual selection both act on the heritable variation among individuals within a population that leads to differential fitness. Sexual selection refers specifically to the differential reproductive success that results because individuals vary in traits that affect their ability to compete for fertilizations. So sexual selection can be considered a specific type of natural selection (see Chapter 9).

   d. Sexual selection does not really differ from natural selection, but the optimum trait value under sexual selection can often be at odds with the optimum trait value under natural selection.

   Correct, but so are other answers. Natural selection and sexual selection both act on the heritable variation among individuals within a population that leads to differential fitness. Sexual selection refers specifically to the differential reproductive success that results because individuals vary in traits that affect their ability to compete for fertilizations. Sexual selection can produce phenotypic traits that actually reduce fitness of many of the individuals that possess them, however (e.g., long, gangly tails that impair flight; bright colors or mating behaviors that attract the attention of predators), especially in populations where the operational sex ratio is highly skewed (see Chapter 9).

   e. Both c and d.

   Correct. Natural selection and sexual selection both act on the heritable variation among individuals within a population that leads to differential fitness. Sexual selection refers specifically to the differential reproductive success that results because individuals vary in traits that affect their ability to compete for fertilizations. In that sense, sexual selection can be considered a specific type of natural selection. Sexual selection also can produce phenotypic traits that actually reduce the fitness of many of the individuals that possess them, however (e.g., long, gangly tails that impair flight; bright colors or mating behaviors that attract the attention of predators), especially in populations where the operational sex ratio is highly skewed (see Chapter 9).

2. What are clades?

   a. Groups made up of organisms and all of their descendants.

   Correct, but so are other answers. A clade is an organism and all its descendants—it is a single branch in a phylogeny (Chapter 4).

   b. Hierarchies nested according to shared traits that are inherited from recent common ancestors.

   Correct, but so are other answers. Clades are hierarchies nested according to traits shared with the most recent common ancestor (Chapter 4).

   c. Groups of organisms that comprise a taxonomic unit.

Incorrect. Taxonomic units do not always align with clades because many well-known taxonomic units were based on different lines of evidence. Recently some taxonomic units, such as reptiles, have been revised because they are not monophyletic clades.

d. Groups of living organisms that share a phenotypic trait or character state.

Incorrect. Not all groups share traits because they are inherited them from a common ancestor. Sometimes, trait similarities arise by convergent evolution. So just because a group of organisms shares a phenotypic trait or character state doesn't make that group a clade. Additional evidence is necessary to know whether a phenotypic trait or character state was inherited from a recent common ancestor (Chapter 8).

e. Both a and b.

Correct. A clade is an organism and all its descendants—it is a single branch in a phylogeny. Clades are "nested" according to the traits shared with the most recent common ancestor (Chapter 4). Clades do not necessarily define taxonomic units, however, because many well-known taxonomic units were based on different lines of evidence. Recently some taxonomic units, such as reptiles, have been revised because of growing evidence relating dinosaurs and birds. Similarly, some organisms share characteristics because of convergent evolution, and these groups do not represent clades.

3. How accurate is the current estimate for the age of the Earth (4.567 billion years)?

a. Not accurate at all because it is just an estimate.

Incorrect. Scientists from a variety of disciplines have been measuring decay rates and incorporating these measurements into radioactive clocks that have been tested, reviewed, and retested. The probabilistic mathematical equations they use based on this information provide very narrow estimates of the ages of geological formations. Scientists have concluded with strong confidence that the Earth began to form from the solar system's primordial dust cloud 4.567 billion years ago (see Box 3.1).

b. Only slightly accurate because radiometric dating requires knowing decay rates, and no one has measured decay rates directly.

Incorrect. Scientists from a variety of disciplines have been measuring decay rates for decades. They've incorporated these measurements into radioactive clocks that have been tested, reviewed, and retested. The probabilistic mathematical equations they use based on this information provide very narrow estimates of the ages of geological formations. Scientists have concluded with strong confidence that the Earth began to form from the solar system's primordial dust cloud 4.567 billion years ago (see Box 3.1).

c. Somewhat accurate, but different dating techniques give different results.

Incorrect. Scientists often test their results using different dating techniques because the weight of evidence is important when making scientific claims. Thousands of research papers are published each year on radiometric dating, essentially all in agreement. Indeed, the fact that they can derive the same age with independent methods confirms that radiometric dating is a valid way to measure the age of rocks. Scientists have concluded with strong confidence that the Earth began to form from the solar system's primordial dust cloud 4.567 billion years ago (see Box 3.1).

d. Fairly accurate because scientists have been replicating the experiments and reevaluating the evidence for decades.

Correct. Scientists often test their results using different dating techniques because the weight of evidence is important when making scientific claims. Thousands of research papers are published each year on radiometric dating, essentially all in agreement. Indeed, the fact that they can derive the same age with independent methods confirms that radiometric dating is a valid way to measure the age of rocks. Scientists have concluded with strong confidence that the Earth began to form from the solar system's primordial dust cloud about 4.567 billion years ago (the margin of error is less than 0.1 percent—that's less than one million years; see Box 3.1).

Identify Key Terms

allopatry	Populations occurring in separate, nonoverlapping geographic areas (i.e., they are separated by geographic barriers to gene flow).
allopolyploidy	The occurrence of more than two paired chromosomes as a result of interspecific hybridization.
biological species concept	Species are groups of actually (or potentially) interbreeding natural populations that are reproductively isolated from other such groups.
cryptic species	Groups of organisms that are genetically distinct and do not interbreed, but are morphologically almost indistinguishable.
geographic barriers	Features of the environment that physically separate populations from each other.
horizontal gene transfer	Any process in which genetic material is transferred to another organism without descent.
isolating barrier	Intrinsic properties of organisms that reduce the likelihood of interbreeding between individuals of different populations.
phylogenetic species concept	Species are the smallest possible groups whose members are descended from a common ancestor and who all possess defining or derived characteristics that distinguish them from other such groups.
reproductive barriers	An aspect of the environment, genetics, behavior, physiology, or ecology of a species that reduces or impedes gene flow from individuals of other species.
reproductive isolation	Occurs when reproductive barriers prevent or strongly limit reproduction between populations. The result is that few or no genes are exchanged between the populations.
ring species	A connected series of populations, each of which can interbreed with its neighboring populations, that has diverged sufficiently across its range that the populations at the ends of the series are too different to interbreed.
speciation	The evolutionary process by which new species arise, resulting in one evolutionary lineage that splits into two or more lineages.
sympatry	Populations occurring in the same geographic area.

Link Concepts

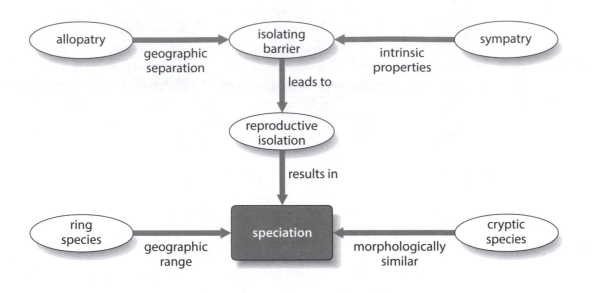

Interpret the Data

- Were *Drosophila* species pairs more likely to hybridize when they had a genetic distance < 0.5 or > 0.5?

  Drosophila species pairs were more likely to hybridize when they had a genetic distance < 0.5. For closely related species (those with a genetic distances less than 0.5), the amount of hybridization ranged from about 0.15 to 1. Some species pairs were hardly isolated from each other at all, and others were completely isolated, reproductively. For species pairs that were more distantly related (those with a genetic distances greater than 0.5), the amount of hybridization ranged from about 0.45 to 1. So in general, species that were more closely related were more likely to hybridize than species that were less closely related—even though all of the species pairs did hybridize at some level.

- Did *Drosophila* species pairs hybridize even if they weren't closely related?

  Yes, even species that were very distantly related (values greater than 1) were not completely isolated reproductively.

- If it takes roughly a million years for D to reach a value of 1, when would you consider *Drosophila* species to be reproductively isolated?

  Drosophila species could be considered reproductively isolated in about ¾ of a million years (D between 0.5 and 1) because by then a typical pair of Drosophila species no longer interbred.

Games and Exercises

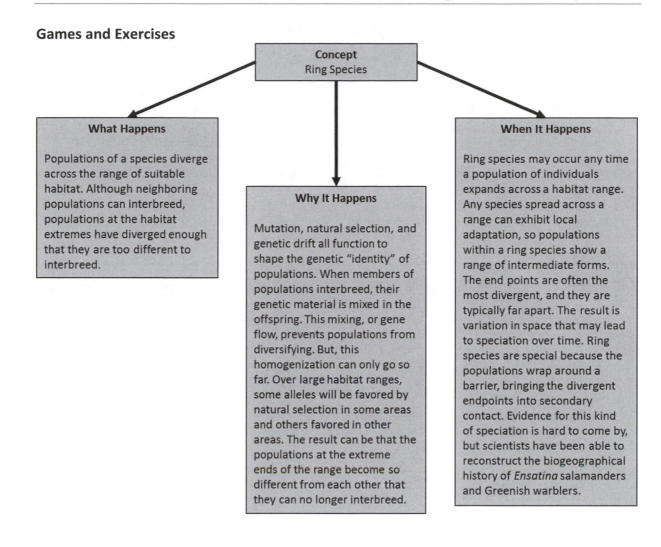

Concept
Ring Species

What Happens

Populations of a species diverge across the range of suitable habitat. Although neighboring populations can interbreed, populations at the habitat extremes have diverged enough that they are too different to interbreed.

Why It Happens

Mutation, natural selection, and genetic drift all function to shape the genetic "identity" of populations. When members of populations interbreed, their genetic material is mixed in the offspring. This mixing, or gene flow, prevents populations from diversifying. But, this homogenization can only go so far. Over large habitat ranges, some alleles will be favored by natural selection in some areas and others favored in other areas. The result can be that the populations at the extreme ends of the range become so different from each other that they can no longer interbreed.

When It Happens

Ring species may occur any time a population of individuals expands across a habitat range. Any species spread across a range can exhibit local adaptation, so populations within a ring species show a range of intermediate forms. The end points are often the most divergent, and they are typically far apart. The result is variation in space that may lead to speciation over time. Ring species are special because the populations wrap around a barrier, bringing the divergent endpoints into secondary contact. Evidence for this kind of speciation is hard to come by, but scientists have been able to reconstruct the biogeographical history of *Ensatina* salamanders and Greenish warblers.

Delve Deeper

1. Do geographic isolating barriers function equally under the phylogenetic and biological species concepts?
   No. Geographic isolating barriers can interrupt gene flow between populations of individuals. These populations may then evolve independently as a result of drift, mutation, and natural selection. According to the phylogenetic species concept, as long as the populations have evolved independently, so that there are recognizable sets of individuals with distinct patterns of ancestry and descent, then the populations can be considered species. The biological species concept, however, would consider the two populations as isolated, but they could still potentially interbreed. Geographic isolation may or may not lead to the evolution of reproductive isolation.

2. Why would different kinds of bacteria and archaea be difficult to identify as different species using the biological species concept?
   Bacteria and archaea reproduce asexually, dividing into genetically identical copies. So reproductive barriers used in the biological species concept are meaningless for them. Genetic and ecological differences might be more useful in classifying bacteria and archaea.

3. Explain how firefly flashes can act as reproductive barriers.

The males have species specific flashes that must be recognized by the females before they respond. The females then wait a species specific length of time before answering with a certain flash pattern that the males must recognize. If either sex doesn't recognize the other as a potential mate, their behavior will act as a barrier to reproduction.

4. Predict what would happen if you took a female greenish warbler from the oldest, extreme southern end of its distribution and transplanted her to the most northern end of the warbler's range in Siberia, where the two extremes of the ring species overlap. Would she be able to successfully mate with either an east or west Siberian warbler? Both? Neither? Explain why referring to both behavior and genetics.

Two results are possible. The first possibility is she would be able to mate with either of the extremes because she is halfway between either extreme in geography and genetics, roughly in the center of the genetic cline. Another possibility is that she couldn't mate with either because too much time had passed since their populations were directly interbreeding with hers. The outcome depends on if she will respond to their behavioral courtship and songs and if they are still genetically compatible.

5. What are the similarities and differences between allopatric and sympatric speciation?

The similarities lie in the fact that both processes lead to the splitting of one species into two. Both processes involve genetic diversification and reproductive isolating mechanisms. The differences are in where these events happen and how those circumstances impact the conditions of speciation. Allopatric speciation happens in geographically separate areas, so potentially reproducing individuals are already isolated.

Sympatric speciation happens in the same geographic location where individuals can come in contact with each other, so some other mechanism (e.g., divergent ecological selection) functions to isolate the populations.

6. Under what ecological and evolutionary conditions is sympatric speciation most likely?

Sympatric speciation is most likely when individuals from the same species and same location are separated by preferences for habitat or resources, keeping them from coming in contact with each other especially during mating. The speciation of Rhagoletis flies could be considered sympatric because the hawthorn trees and apple trees exist in the same field (they are not geographically isolated).

7. Which species concept would you apply to plants that hybridize relatively frequently, such as salsify flowers (*Tragopogon*), often resulting in allopolyploidy?

Because the biological species concept identifies species as groups of actually, or potentially, interbreeding populations, it may not be universally applicable to plants. Clearly, reproductive isolating barriers break down or do not exist among many easily identifiable species (to humans). Several known cases of *Tragopogon* speciation have resulted from hybridization and allopolyploidy. The phylogenetic species concept may be more appropriate when studying plant speciation, especially speciation due to allopolyploidy, because groups descended from the common ancestor whose members all possess the defining or derived characteristics can be easily distinguished from other groups.

Test Yourself

1. e; 2. d; 3. c; 4. c; 5. a; 6. b; 7. a; 8. e; 9. b; 10. d

Chapter 11
MACROEVOLUTION—LIFE OVER THE LONG RUN

Check Your Understanding

1. What is a species?
   a. A distinct kind of organism that is easily identifiable.
   b. The smallest possible group whose members are descended from a common ancestor and who all possess defining or derived characteristics that distinguish them from other such groups.
   c. A group of actually or potentially interbreeding natural populations that is reproductively isolated from other such groups.
   d. All of the above.
   e. b and c only.

2. Why don't scientists agree on a single definition of species?
   a. Because they have not discovered the true definition of a species.
   b. Because research methods can dictate which definition is most useful.
   c. Because different scientists have different philosophies about defining species.
   d. All of the above.
   e. b and c only.

3. Why isn't the fossil record a complete record of life on earth?
   a. Because organisms eat other organisms.
   b. Because conditions have to be just right in order to preserve fossils.
   c. Because wind and rain can erode fossils from the substrate.
   d. Because fossil-bearing rocks can be difficult to access.
   e. All of the above.
   f. a and b only.
   g. c and d only.

Learning Objectives for Chapter 11:

➢ Compare and contrast macroevolution and microevolution in terms of processes and patterns.
➢ Evaluate the evidence supporting vicariance and dispersal events in the evolution of marsupials.
➢ Describe the central insight of punctuated equilibria.
➢ Explain when adaptive radiations can take place.
➢ Compare background extinctions with mass extinctions and define mass extinction.
➢ Describe two factors potentially responsible for mass extinctions.
➢ Evaluate the evidence for how the actions of humans may lead to another mass extinction.

Identify Key Terms

Connect the following terms with their definitions on the right:

adaptive radiations	The variety of life, including the numbers or abundance of organisms within a habitat, an ecosystem, or even the entire planet; the variety of ecological and biological interactions; and the range of genetic differences.
background extinction	The evolutionary process by which new taxa arise.
biodiversity	That point when the last member of a clade dies; the clade may be a single species or a higher group.
biogeography	A single "branch" in the tree of life representing an organism and all of its descendants.
Cambrian Explosion	The study of the distribution of species across space (geography) and time.
clade	The formation of geographic barriers to dispersal and gene flow, resulting in the separation of populations.
dispersal	The movement of populations from one geographic region to another.
extinction	A group of mammals found principally in the Southern Hemisphere that does not develop true placentas. Females usually have a pouch on their abdomen that functions to carry the young.
hypothesis	A model of evolution that proposes that most species undergo relatively little change for most of their geologic history. These periods of stasis are punctuated by brief periods of rapid change associated with speciation events.
macroevolution	Evolutionary clades that have undergone exceptionally rapid diversification into a variety of lifestyles or ecological niches.
marsupial	An extraordinary period of evolution in the animal kingdom that occurred over a relatively short period of time around 540 million years ago.
mass extinction	A proposed explanation for a natural phenomenon that is based in evidence and can be tested with additional evidence.
microevolution	A statistically significant departure from background extinction rates that results in a substantial loss of taxonomic diversity.
origination	Evolution occurring above the species level, including the origination, diversification, and extinction of species over long periods of evolutionary time.
punctuated equilibria	Evolution occurring within populations, including adaptive and neutral changes in allele frequencies from one generation to the next.
vicariance	The normal rate of extinction for a taxon or collection of organisms.

Link Concepts

1. Fill in the following concept map with the key terms from the chapter

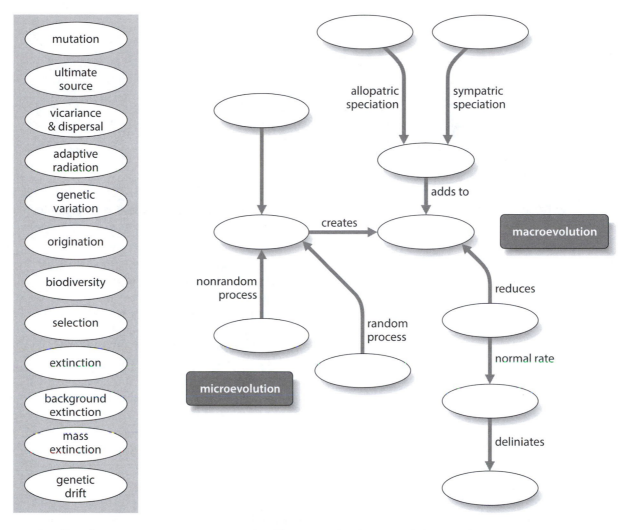

mutation

ultimate source

vicariance & dispersal

adaptive radiation

genetic variation

origination

biodiversity

selection

extinction

background extinction

mass extinction

genetic drift

allopatric speciation

sympatric speciation

adds to

creates

macroevolution

nonrandom process

random process

reduces

microevolution

normal rate

deliniates

2. Develop your own concept map that illustrates vicariance and dispersal events in the evolution of marsupials. Draw circles representing Africa, Antarctica, Asia, Australia, Europe, North America, and South America. Using different colored pens, connect the dispersal events, then use big red "X"s to add the vicariance and extinction events.

Interpret the Data

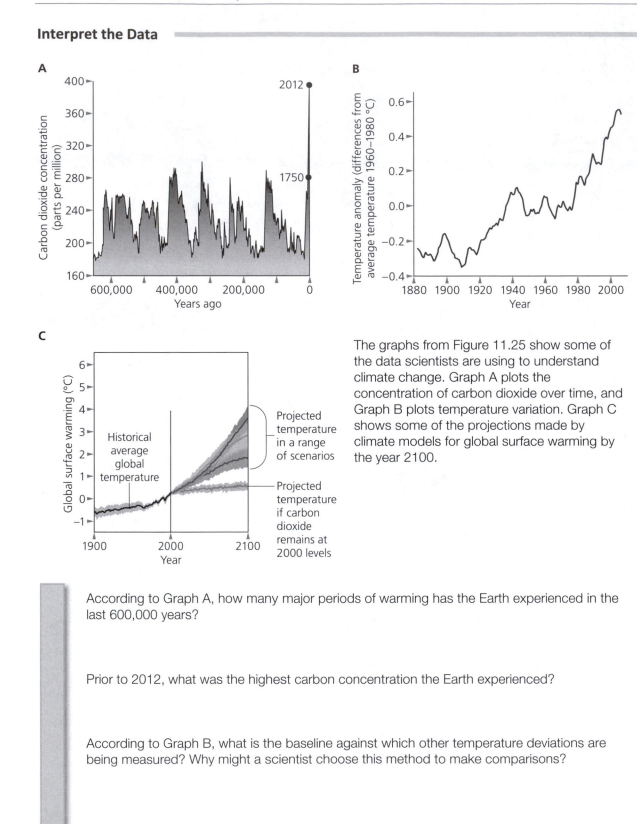

A

Carbon dioxide concentration (parts per million)

2012

1750

600,000   400,000   200,000   0

Years ago

B

Temperature anomaly (differences from average temperature 1960–1980 °C)

1880 1900 1920 1940 1960 1980 2000

Year

C

Global surface warming (°C)

Historical average global temperature

Projected temperature in a range of scenarios

Projected temperature if carbon dioxide remains at 2000 levels

1900   2000   2100

Year

The graphs from Figure 11.25 show some of the data scientists are using to understand climate change. Graph A plots the concentration of carbon dioxide over time, and Graph B plots temperature variation. Graph C shows some of the projections made by climate models for global surface warming by the year 2100.

According to Graph A, how many major periods of warming has the Earth experienced in the last 600,000 years?

Prior to 2012, what was the highest carbon concentration the Earth experienced?

According to Graph B, what is the baseline against which other temperature deviations are being measured? Why might a scientist choose this method to make comparisons?

Does the variation in projected temperatures for the year 2100 mean that scientists don't know what they are talking about?

Games and Exercises

How Big Is a Billion?

We live in very short time frames (for example, the original iPhone was released in 2007!). As a result, the "deep time" over which evolution has taken place can be exceptionally difficult to comprehend. Here's a quick visual trick that can help you grasp exactly how long a billion years is. All you need is 10 quarters (100 quarters would be awesome!), a ruler, and a calculator. Use your ruler to determine how tall a quarter is when it is lying flat on a table (you can use English or metric, whichever is more comfortable to you). Stack the 10 quarters, one on top of another in a nice, neat stack. Use your ruler to determine how tall the stack is—it should be 10 times as tall as a single quarter. (If you have 100 quarters, see if you can stack them and measure how tall that stack is.) Now use your calculator to fill in this table:

# of quarters	height
1	
10	
100	
1000	
10,000	
100,000	
1,000,000	
10,000,000	
100,000,000	
1,000,000,000	

How tall are you? (60 inches = 5 feet or 152 cm)

About 1/10 of a mile or 1 ¾ football fields (0.16 km)

How far do you live from your school? (63,360 inches = 1 mile or 1.61 km)

How far do you live from Anchorage, Alaska?

Adaptive Radiation in the Hawaiian Honeycreepers

Hawaiian Honeycreepers are an amazing group of birds. The ancestors of this group first colonized the volcanic islands of Hawai'i about 5 million years ago. They diversified into dramatically different forms—identifiable largely by their diverse bills and feeding habits (Figure 11.11). To get a feel for how natural selection could act so rapidly to lead to such different forms, you can do a little experiment with bill shapes and foraging efficiency.

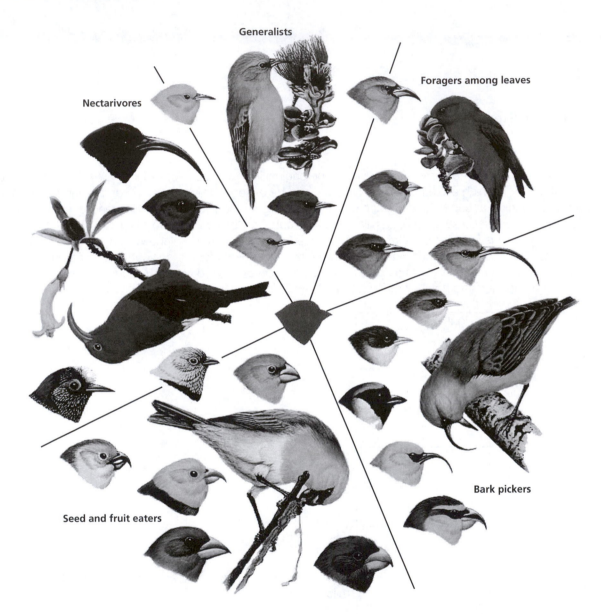

Adapted from Losos and Ricklefs (2009).

You will need a pair of regular pliers, a pair of needle-nosed pliers, a pair of scissors, some peanuts or walnuts in the shell, sunflower seeds (in the shell), uncooked rice, a bottle of water (make sure the neck is somewhat narrow), a bowl, and a small glass. The nuts, seeds, rice, and water represent food types. Place some nuts, seeds, and rice on a flat table and open the bottle of water and place the glass nearby. Use each tool with each type of food and see how easy it is to "eat" by transferring the food from the table to the bowl (or small glass for the water). Time your trials. Give yourself one minute with each, and fill out the data table below. Make it harder by actually cracking the nuts or seeds.

Beak Type	# peanuts or walnuts	# sunflower seeds	# rice	height of water in glass
regular pliers				
needle-nosed pliers				
tweezers				
medicine dropper				

- Did any single beak type work best for all foods? If not, which type worked best for which foods?

- How could different beak types evolve from a common ancestor through natural selection?

- If the Hawaiian Islands differed radically in the types of habitats available (say islands to the west are desert-like and islands to the east are rainforest-like), would you predict that different groups of birds would exist on different islands? Why?

Now look at some specific honeycreepers. See if you can match their bill types with the types of food they can access.

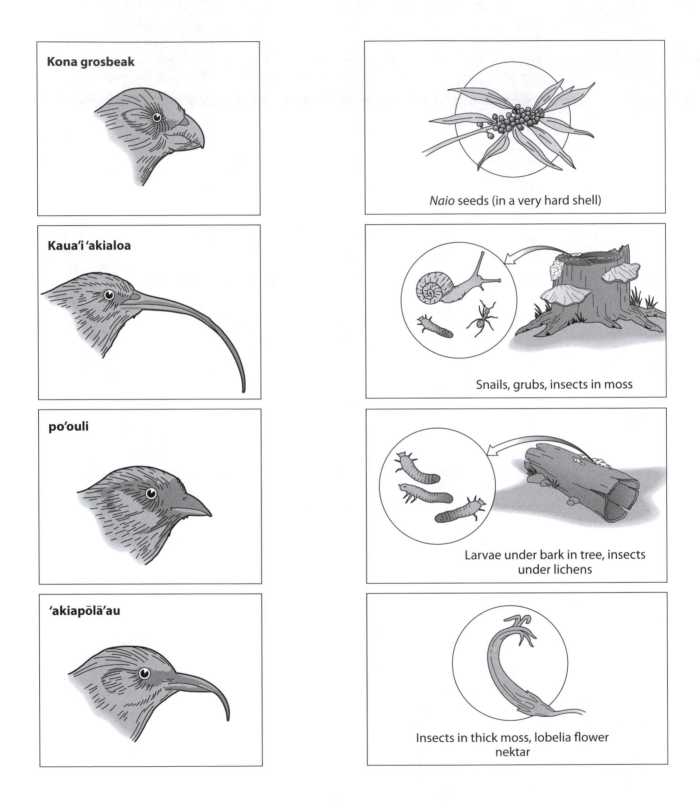

Kona grosbeak

Kaua'i 'akialoa

po'ouli

'akiapōlā'au

Naio seeds (in a very hard shell)

Snails, grubs, insects in moss

Larvae under bark in tree, insects under lichens

Insects in thick moss, lobelia flower nektar

These four bill forms represent some of the diversity in the honeycreeper radiation.

➤ Kona grosbeaks used their thick bills to crack open hard seeds, such as the naio. Unfortunately, the Kona grosbeak went extinct, even though the naio is quite abundant.

➤ The Kaua'i 'akialoa foraged for insects that lived in deep mats of moss. Its unusually long bill also allowed it to access nectar from long tubular flowers, like lobelia. Prior to their extinction, different species of 'akialoa (and two fossil species) were found on different islands in the Hawaiian chain, including Kaua'i.

➤ The po'ouli used its small bill to feed on snails, grubs, and insects. The po'ouli is likely extinct as well—the last known birds were observed on Maui in 2004.

➤ The Akiapola'au still inhabits the island of Hawai'i, but its numbers are plummeting. Like a woodpecker, it uses its sharp beak to forage for insects.

The Hawaiian Islands offered a diverse, but fragile, environment for this extraordinary radiation of birds, but they are a cautionary tale. Humans introduced predators and mosquito-borne diseases to the island, and also destroyed Hawai'i's pristine habitat. As a result, less than half of the species first discovered there still remain.

Adapted from ERUPTION, an electronic field trip produced by Ball State University (http://electronicfieldtrip.org/volcanoes/teachers/classroom_honeycreeper.html).

Go Online

100 Greatest Discoveries: Continental Drift

Alfred Wegener was a meteorologist and geophysicist. This short clip from How Stuff Works illustrates some of the biological evidence Wegener used when he developed his ideas about continental drift.

http://geography.howstuffworks.com/29267-100-greatest-discoveries-continental-drift-video.htm

Earth's Paleogeography—Continental Movements through Time

This cool animation created by Dr. Ron Blakey at Northern Arizona University shows the history of the continents on Earth and points out when the major radiations and extinctions of organisms occurred—you can see what the land masses of the continents looked like when these major changes in biodiversity were occurring!

http://www.youtube.com/watch?v=GNmUd43pabg

Mass Extinctions

Hank Green explains the Big Five mass extinctions and offers clues to what could be the sixth.

http://www.youtube.com/watch?v=FlUes_NPa6M

Mass Extinction

In this edition of *NOVA Science* Now, Neil deGrasse Tyson explores the causes of the Permian Mass Extinction 250 million years ago and its effects on life on Earth.

http://www.pbs.org/wgbh/nova/earth/mass-extinction.html

Facts of Evolution: Speciation and Extinction

In this comprehensive video, Best of Science and the Science Channel use the fossil record to show that extinction has always been a part of the Earth's history, but so has speciation. More importantly, they use current evidence to show how speciation arises and the direct and indirect evidence for the evolution of the diversity of life.

http://www.youtube.com/watch?v=T5kumHLiK4A

Scientists Gauge Ancient Die-Off of Pacific Birds

Humans have often been the cause of massive extinctions. For example, the Pacific islands were home to a huge diversity of birds before the arrival of humans, but the total loss couldn't be determined because of the incomplete nature of the fossil record. Now a new study estimates that nearly 1000 species went extinct—about 1/10th of the world's species.

http://news.sciencemag.org/plants-animals/2013/03/scientists-gauge-ancient-die-pacific-birds?ref=em

Ancestor of All Placental Mammals Revealed

Aided by the wealth of mammal fossils that have been discovered and a phylogeny based on anatomy, physiology, and genetics, scientists have inferred what the common ancestor of placental mammals probably looked like.

http://news.sciencemag.org/sciencenow/2013/02/ancestor-of-all-placental-mammal.html?ref=em

Gut Microbes Can Split a Species

For a whole new take on speciation, scientists examined the microbes harbored in the guts of organisms to see if differences in biodiversity at that scale affect speciation at the larger scale.

http://news.sciencemag.org/evolution/2013/07/gut-microbes-can-split-species

Beaks Keep Avian Lovers Apart

 And now for an indirect effect of natural selection on bill size and shape—morphology influences sexual selection and speciation!

http://news.sciencemag.org/2001/01/beaks-keep-avian-lovers-apart

Science*Shot: Big Smash, Dead Dinos*

Did dinosaurs go extinct because of an asteroid or was the asteroid just the coup de grâce (the death blow) to a group that was fading fast? New research using a high-resolution radiometric dating technique provides some pretty convincing evidence for one of the hypotheses.

http://news.sciencemag.org/2013/02/scienceshot-big-smash-dead-dinos?ref=em

Let There Be Mammals

When did the radiation of placental mammals occur? Did it occur only after the disappearance of the dinosaurs, or did it begin while dinosaurs were still around? This research paper from Science presents some striking new evidence that living placental mammals originated and radiated after the Cretaceous.

http://app.aaas-science.org/e/er?s=1906&lid=26839&elq=c49daf2845264f4b919e607a3ffc81e7

Or read the *Science Perspective* by Anne D. Yoder:

 http://app.aaas-science.org/e/er?s=1906&lid=26798&elq=c49daf2845264f4b919e607a3ffc81e7

Common Misconceptions

Microevolution and Macroevolution Are Two Sides of the Same Coin

Microevolution and macroevolution can be distinguished by the patterns they describe, but underneath it all, the mechanisms leading to those patterns are the same. Microevolution describes the genetic variation within populations, and how that variation came about through mutation, selection, and drift. Microevolution, or changes in allele frequencies over time, has been well documented, and given enough time, microevolutionary processes can produce enough differences between populations that they can be considered separate species.

Macroevolution, on the other hand, describes origination, diversification, and extinction of groups of organisms over time. The periods of time necessary for macroevolutionary patterns to appear can be unfathomably long, especially when our own experiences are usually just measured in decades. Although such long periods of time make direct observation of these changes unlikely, macroevolutionary patterns are very apparent in the fossil record. More importantly, volumes of indirect evidence, especially in the burgeoning field of evolutionary development, indicate that given enough time, microevolutionary changes can accumulate and translate into macroevolutionary change.

Gaps in the Fossil Record and Evolution
The fossil record is not complete, and we can't ever expect it to be. As Chapter 3 illustrates, fossilization is a complex process, and fossilization events that preserve the diversity of life, even as just snapshots in time, are extremely rare. When Darwin introduced natural selection as a process for producing the diversity of organisms on Earth, scientists were just beginning to be able to predict fossil locations. As a component of evolutionary

theory, natural selection and common ancestry gave scientists an additional framework for making predictions about where and when to locate transitional fossils.

In fact, since Darwin's time, scientists have discovered many transitional fossils (including transitional fossils within the human lineage, between land mammals and modern whales, and between dinosaurs and modern birds). But because of the nature of the fossil record, they don't expect to find every transition. And more importantly, missing transitional fossils doesn't negate the overwhelming evidence for evolution.

Contemplate

Based on your understanding of vicariance and dispersal, what might you predict about the relationships between South American plants and African plants (you might want to use Figure 11.7 as a guide)?	Could the central insights scientists gained from punctuated equilibria be applied to evolution of the English language? How?

Delve Deeper

1. What are the differences between microevolution and macroevolution? What are the similarities? Which has more evidentiary support, microevolution or macroevolution?

2. What lines of evidence have macroevolutionary biologists used to determine the origin of marsupials? How are these lines distinct?

3. Are punctuated equilibria and Darwin's theory of natural selection at odds in evolutionary theory?

4. How is the term Cambrian Explosion misleading?

5. Should humans be concerned about the pace of extinctions of organisms that are not directly related to our survival?

Go the Distance: Examine the Primary Literature

Peter Mayhew and his colleagues examine biodiversity patterns in marine invertebrates during the Phanerozoic Eon (an eon is a time period longer than an era; the Phanerozoic Eon includes the Paleozoic, Mesozoic, and Cenozoic eras). An important component of their research included controlling for the bias historically inherent in paleontological samples.

- According to the authors, what causes bias in studies of paleodiversity?

- Why did Mayhew and his colleagues argue that this sampling bias must be considered?

- What did Mayhew and his colleagues find?

Mayhew, P. J., M. A. Bell, T. G. Benton, and A. J. McGowan. 2012. Biodiversity Tracks Temperature over Time. *Proceedings of the National Academy of Sciences* 109: 15141–45. http://www.pnas.org/content/early/2012/08/27/1200844109.abstract.

Test Yourself

1. How are originations and extinctions related to biodiversity?
   a. Biodiversity is a result of the relationship between origination rates and extinction rates.
   b. Originations and extinctions are less important to biodiversity within a specific geographic area than other processes, such as immigration and emigration.
   c. Originations always lead to an increase in biodiversity and extinctions always lead to a decrease in biodiversity.
   d. Both originations and extinctions have increased over time.
   e. Both c and d explain the relationship between extinctions and biodiversity.

2. Which of these statements about vicariance is TRUE?
   a. Plate tectonics are a primary mechanism of vicariance.
   b. Vicariance led to Australian and South American lineages of marsupials.
   c. Vicariance prevents dispersal.
   d. Both a and b are true.
   e. All of the above are true.

3. Punctuated equilibria is a hypothesis about the evolution of species. According to this hypothesis, why should macroevolutionary biologists expect abrupt breaks in the fossil record?
   a. Because the processes that form fossils are inconsistent through time, so there will always be gaps in the record.
   b. Because significant changes in phenotypes may occur within small isolated populations on the fringes of a species' range, leading to rapid speciation and little fossil evidence.
   c. Because species are defined as substantial morphological shifts within the fossil record.
   d. Macroevolutionary biologists should not expect abrupt breaks in the fossil record because entire populations most often transform from one form to another.
   e. Macroevolutionary biologists should not expect abrupt breaks in the fossil record. They should expect to find enough fossils to demonstrate a smooth transition from one species to another over time.

4. Which of the following is NOT a hypothesis about the conditions that can lead to adaptive radiations?
   a. Adaptive radiations occur as a result of the emergence of new habitats.
   b. Adaptive radiations occur as a result of island formation.
   c. Adaptive radiations occur as a result of the emergence of ecological opportunities.
   d. All are hypotheses about the conditions that can lead to adaptive radiations.
   e. None is a hypothesis about the conditions that can lead to adaptive radiations.

5. Which of the following statements about the Cambrian Explosion is FALSE?
   a. Changing oxygen levels in the Earth's oceans may have created ecological opportunities that organisms were able to exploit.
   b. The emergence of the genetic toolkit may have allowed the evolution of new developmental pathways leading to a diversity of body forms.
   c. The Cambrian Explosion was not really an explosion because the complex body plans of animal taxa evolved from precursors—they did not simply appear.
   d. The sudden appearance of sponges marks the beginning of the Cambrian Explosion.
   e. Molecular clocks indicate that the common ancestor of all animals lived about 800 million years ago.

6. The typical tempo of extinctions within a particular taxon is called
   a. Background extinction.
   b. Mass extinction.
   c. The natural extinction rate.
   d. Total extinction.
   e. Episodic extinction.

7. How are mass extinctions defined?
   a. As a statistical departure from background extinction rates.
   b. As any large number of species that disappears at a point in the stratigraphic column.
   c. As the extinction of many species in a short period of time.
   d. As the extinction of many species as a result of massive volcanic eruptions.

8. Can the Big Five extinctions all be attributed to a single cause? If so, what caused them?
   a. Yes. The Big Five extinctions were caused by asteroids that had major impacts on habitats when they hit the Earth.
   b. Yes. The Big Five extinctions resulted from plate tectonics that changed the quantity and quality of available habitats.
   c. No. The Big Five extinctions were caused by various factors that affected different taxa differently.
   d. No. The Big Five extinctions resulted from low origination rates that resulted from a variety of biotic factors.
   e. No. The Big Five extinctions are statistical anomalies caused by examining families as taxonomic units instead of species.

9. Why don't all scientists accept an extraterrestrial explanation for the mass extinction at the end of the Cretaceous period?
   a. Because losses in marine invertebrates started several million years before the end of the Cretaceous.
   b. Because the disappearance of some taxa may have been the result of volcanic activity that began before evidence of the asteroid impact appeared.
   c. Because extinctions among many terrestrial animals were not that distinct from background rates.
   d. Because the diversity of some dinosaur clades had been declining for several million years before the asteroid impact.
   e. All of the above.

10. Which is not a factor contributing to the looming sixth mass extinction?
   a. Human-caused increases in carbon dioxide.
   b. Collapse of food webs.
   c. Human-caused extinctions.
   d. Habitat loss.
   e. Volcanic eruptions.

Answers for Chapter 11

Check Your Understanding

1. What is a species?
   a. A distinct kind of organism that is easily identifiable.

   Incorrect. This definition does not provide any useful way of distinguishing groups of organisms, especially taxonomic levels. A distinct "kind" could be fungi (a kingdom) or mammals (a class), a duck (an order), or a duck-billed platypus (a species). In addition, our senses are limited, and we can't necessarily observe differences—species may not be easily identifiable to us. Cryptic species are defined as groups of organisms that are genetically distinct and do not interbreed but are morphologically almost indistinguishable. Other groups of organisms can be widely distributed, and although neighboring populations can interbreed, populations at either end cannot. Some identifiable groups of organisms reproduce asexually, so concepts related to interbreeding don't necessarily apply, and other identifiable groups readily interbreed, even though they remain easily identifiable (see Chapter 10).

   Our human nature pushes us to classify things into separate and distinct groups, but nature doesn't necessarily work that way. Groups of organisms clearly share characteristics at different levels—species, genera, families, orders, and so on. Scientists may not be able to come up with a single definition for species, but they understand that that kind of classification is a human artifact. "Species" doesn't have to be defined precisely—unequivocally—for evolutionary theory to explain the existence of groups of organisms that share characteristics with a common ancestor. The National Center for Science Education has a great essay explaining species concepts (http://ncse.com/rncse/26/4/species-kinds-evolution).

   b. The smallest possible group whose members are descended from a common ancestor and who all possess defining or derived characteristics that distinguish them from other such groups.

   Correct, but so are other answers. This definition is called the phylogenetic species concept, a concept used by scientists examining evolutionary lineages (see Chapter 10). Cladogenesis is the event when one evolutionary group (lineage) splits into two or more groups (lineages). Our human nature pushes us to classify things into separate and distinct groups, but nature doesn't necessarily work that way. Defining exactly what constitutes a species is difficult, but scientists understand that that kind of classification is a human artifact. "Species" doesn't have to be defined precisely—unequivocally—for evolutionary theory to explain the existence of groups of organisms that share characteristics with a common ancestor. The National Center for Science Education has a great essay explaining species concepts (http://ncse.com/rncse/26/4/species-kinds-evolution).

   c. A group of actually or potentially interbreeding natural populations that is reproductively isolated from other such groups.

   Correct, but so are other answers. This definition is called the biological species concept. It can be very useful to scientists studying organisms that reproduce sexually (see Chapter 10). It implies that species can be determined by whether they interbreed or not, and speciation occurs when one lineage splits into two or more lineages, also known as cladogenesis. Our human nature pushes us to classify things into separate and distinct groups, but nature doesn't necessarily work that way. Defining exactly what constitutes a species is difficult, but scientists understand that that kind of classification is a human artifact. "Species" doesn't have to be defined precisely—unequivocally—for evolutionary theory to explain the existence of groups of organisms that share characteristics with a common ancestor. The National Center for Science Education has a great essay explaining species concepts (http://ncse.com/rncse/26/4/species-kinds-evolution).

   d. All of the above.

   Incorrect. Defining a species as a "distinct kind" does not provide any useful way of distinguishing groups of organisms, especially taxonomic levels. A "kind" could be fungi (a kingdom) or mammals (a class), a duck (an order), or a duck-billed platypus (a species). Species can be defined according to the phylogenetic species concept or the biological species concept (or a number of other ways), however. But defining exactly what constitutes a species is difficult, and scientists understand that such classification is a human artifact. "Species" doesn't have to be defined precisely for evolutionary theory to explain the existence of groups of organisms that share characteristics with a common ancestor (see Chapter 10). The National Center for Science Education has a great essay explaining species concepts (http://ncse.com/rncse/26/4/species-kinds-evolution).

e. b and c only.

Correct. Species can be defined according to the phylogenetic species concept or the biological species concept (or a number of other ways). Our human nature pushes us to classify things into separate and distinct groups, but nature doesn't necessarily work that way. Defining exactly what constitutes a species is difficult, and scientists understand that such classification is a human artifact. "Species" doesn't have to be defined precisely for evolutionary theory to explain the existence of groups of organisms that share characteristics with a common ancestor (see Chapter 10). The National Center for Science Education has a great essay explaining species concepts (http://ncse.com/rncse/26/4/species-kinds-evolution).

2. Why don't scientists agree on a single definition of species?

a. Because they have not discovered the true definition of a species.

Incorrect. Scientists don't expect to find a "true" definition of species. Indeed, a single definition for species that covers all taxa may not be possible. Scientists generally agree that there is something special about species, however. Alleles flow among populations within a species differently than they flow between species—different species behave like independent evolutionary units, following separate trajectories (see Chapter 10).

b. Because research methods can dictate which definition is most useful.

Correct, but so are other answers. Defining a species as a unit is an artifact of our human need to classify, but scientists generally agree that there is something special about species. Alleles flow among populations within a species differently than they flow between species—different species behave like independent evolutionary units, following separate trajectories. The precise definition of a species, however, is not clear, and research methods often dictate which definition is most useful. For example, paleontologists deal with morphological differences, and molecular phylogeneticists deal with genetic differences (see Chapter 10).

Species are definitely a concept, but scientists may not agree on how to precisely define a species. The definition is less important than how species come about, however. As John Wilkins points out (http://ncse.com/rncse/26/4/species-kinds-evolution), scientists may not be able to precisely define "mountains," but that doesn't stop their important research on how mountains come about.

c. Because different scientists have different philosophies about defining species.

Correct, but so are other answers. Defining a species as a unit is an artifact of our human need to classify, but scientists generally agree that there is something special about species. Alleles flow among populations within a species differently than they flow between species—different species behave like independent evolutionary units, following separate trajectories. The precise definition of a species, however, is not clear, and different scientists have different philosophies about defining species. For example, some scientists may not believe that species is the most important taxonomic unit of concern (see Chapter 10).

d. All of the above.

Incorrect. Defining a species as a unit is an artifact of our human need to classify, but scientists generally agree that there is something special about species. Alleles flow among populations within a species differently than they flow between species—different species behave like independent evolutionary units, following separate trajectories. There may not be a "true" definition. Research methods often dictate which definition is most useful. Nor do scientists necessarily share the same philosophies about defining species (see Chapter 10).

e. **b and c only.**
Correct. Defining a species as a unit is an artifact of our human need to classify, but scientists generally agree that there is something special about species. Alleles flow among populations within a species differently than they flow between species—different species behave like independent evolutionary units, following separate trajectories. Research methods often dictate which definition is most useful. Nor do scientists necessarily share the same philosophies about defining species. However, there may not be a "true" definition (see Chapter 10).

3. Why isn't the fossil record a complete record of life on earth?

a. **Because organisms eat other organisms.**
Correct, but so are other answers. A very big part of the reason the fossil record is incomplete stems from the fact that organisms eat other organisms, scattering remains and destroying evidence of their existence, let alone leaving anything to fossilize (see Chapter 3).

b. **Because conditions have to be just right in order to preserve fossils.**
Correct, but so are other answers. The conditions have to be just right for an organism to fossilize, many physical processes (wind, waves, running water) and biological processes (fungi, algae) can destroy remains of organisms before they can mineralize. It takes an even rarer set of circumstances for soft tissues, such as skin, to fossilize (see Chapter 3).

c. **Because wind and rain can erode fossils from the substrate.**
Correct, but so are other answers. The same processes that prevent organisms from fossilizing can also destroy fossils that have formed. Substrates, such as sandstones and other sedimentary rocks containing fossils, can become exposed through uplift or erosion. Physical processes, such as wind and rain, can quickly destroy millions of years of mineralization (see Chapter 3).

d. **Because fossil-bearing rocks can be difficult to access.**
Correct, but so are other answers. Many fossils may simply be buried far below the surface currently, or they may be buried under snow and ice, or ocean sediments. Scientists have a limited understanding of the diversity and extent of rock strata, but they have been very successful identifying rocks that should bear fossils (Chapter 4).

e. **All of the above.**
Correct. A very big part of the reason the fossil record is incomplete stems from the fact that organisms eat other organisms, scattering remains and destroying evidence of their existence, let alone leaving anything to fossilize. On the off chance that nothing scavenges the dead organism, the conditions have to be just right for fossilization. Many physical processes (wind, waves, running water) and biological processes (fungi, algae) can destroy remains of organisms before they can mineralize. It takes an even rarer set of circumstances for soft tissues, such as skin, to fossilize. The same processes that prevent organisms from fossilizing can also destroy fossils that have formed. Substrates, such as sandstones and other sedimentary rocks containing fossils, can become exposed through uplift or erosion. Physical processes, such as wind and rain, can quickly destroy millions of years of mineralization. But ultimately, many fossils may simply be buried far below the surface, or they may be buried under snow and ice, or ocean sediments (see Chapter 3).

f. **a and b only.**
Incorrect. A very big part of the reason the fossil record is incomplete stems from the fact that organisms eat other organisms, scattering remains and destroying evidence of their existence, let alone leaving anything to fossilize. And, on the off chance nothing scavenges the dead organism, the conditions have to be just right for fossilization. However, the same processes that prevent organisms from fossilizing can also destroy fossils that have formed, and other fossils may simply be buried far below the surface, under snow and ice, or beneath ocean sediments (see Chapter 3).

g. **c and d only.**
Incorrect. The same processes that prevent organisms from fossilizing can also destroy fossils that have formed, and other fossils may simply be buried far below the surface, under

snow and ice, or beneath ocean sediments. However, a very big part of the reason the fossil record is incomplete stems from the fact that organisms eat other organisms, scattering remains and destroying evidence of their existence, let alone leaving anything to fossilize. And, on the off chance nothing scavenges the dead organism, the conditions have to be just right for fossilization (see Chapter 3).

Identify Key Terms

adaptive radiations	Evolutionary clades that have undergone exceptionally rapid diversification into a variety of lifestyles or ecological niches.
background extinction	The normal rate of extinction for a taxon or collection of organisms.
biodiversity	The variety of life, including the numbers or abundance of organisms within a habitat, an ecosystem, or even the entire planet; the variety of ecological and biological interactions; and the range of genetic differences.
biogeography	The study of the distribution of species across space (geography) and time.
Cambrian Explosion	An extraordinary period of evolution in the animal kingdom that occurred over a relatively short period of time around 540 million years ago.
clade	A single "branch" in the tree of life representing an organism and all of its descendants.
dispersal	The movement of populations from one geographic region to another.
extinction	That point when the last member of a clade dies; the clade may be a single species or a higher group.
hypothesis	A proposed explanation for a natural phenomenon that is based in evidence and can be tested with additional evidence.
macroevolution	Evolution occurring above the species level, including the origination, diversification, and extinction of species over long periods of evolutionary time.
marsupial	A group of mammals found principally in the Southern Hemisphere that does not develop true placentas. Females usually have a pouch on their abdomen that functions to carry the young.
mass extinction	A statistically significant departure from background extinction rates that results in a substantial loss of taxonomic diversity.
microevolution	Evolution occurring within populations, including adaptive and neutral changes in allele frequencies from one generation to the next.
origination	The evolutionary process by which new taxa arise.
punctuated equilibria	A model of evolution that proposes that most species undergo relatively little change for most of their geologic history. These periods of stasis are punctuated by brief periods of rapid change associated with speciation events.
vicariance	The formation of geographic barriers to dispersal and gene flow, resulting in the separation of populations.

Link Concepts

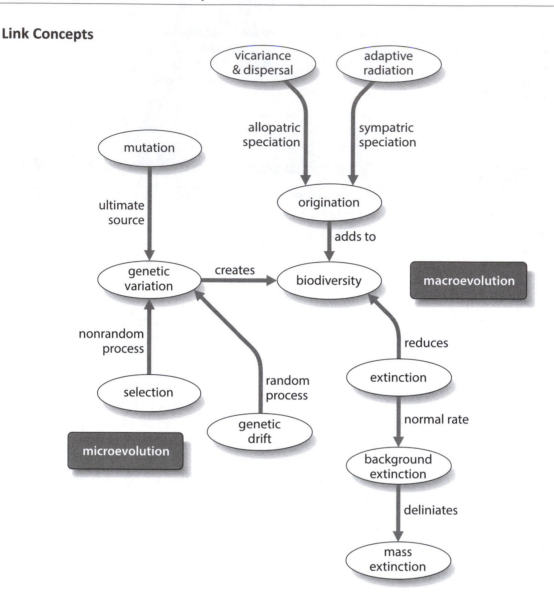

Interpret the Data

- According to Graph A, how many times has the Earth experienced warming in the last 600,000 years?
  Seven.
- Prior to 2012, what was the highest carbon concentration the Earth experienced?
  Approximately 300 parts per million.
- According to Graph B, what is the baseline against which other temperature deviations are being measured? Why might a scientist choose this method to make comparisons?
  The baseline is the average temperature measured over 20 years, from 1960 to 1980. Temperature anomalies are shown as the difference between that baseline and the average temperature for a given year.
  Establishing a baseline allows scientists to understand how observational data may change over time. The baseline establishes a point of reference, and calculating differences from the baseline can help determine relative changes—to understand trends.
  Average annual temperatures are one line of evidence for global climate change, and calculating the statistic—average annual temperature—requires actual measurement of temperatures. (Scientists can also determine estimates of temperature through indirect

measures.) Data on global temperatures can be limiting because of our history and ability to measure global temperature from a diversity of locations. Turns out, the period from 1960 to 1980 has generally better coverage and data availability than earlier years (even though data from earlier years might have been less affected by anthropogenic changes than later years). Calculating differences from this baseline shows that the years before 1960 were relatively cooler and the years after 1980 were relatively warmer. More importantly, the extent of the differences is readily apparent.

- Does the variation in projected temperatures for the year 2100 mean that scientists don't know what they are talking about?

    No. Models take into account a diversity of variables, conditions, and scenarios. Different scientists weigh the evidence that supports those variables differently, so variation is expected. Indeed, having different models with different variables and different conditions is an important part of the scientific process, allowing for rigorous debate and consensus building. The computer models used to make predictions about temperature in the year 2100 consistently show that the planet will warm—even if carbon levels remain at levels measured in the year 2000.

Delve Deeper

1. What are the differences between microevolution and macroevolution? What are the similarities? Which has more evidentiary support, microevolution or macroevolution?

    Microevolution and macroevolution are different emphases along the continuum that is the theory of evolution. Microevolution is the study of small-scale changes, such as changes in allele frequencies, that are affected by selection and drift and their effects on the phenotypes within a population or species. Macroevolution is the study of the large-scale changes that are outcomes of these small-scale processes within the landscape over long periods of evolutionary time. It refers to any evolutionary change at or above the level of species.

    Microevolution and macroevolution have both been observed, but the two emphases focus on different lines of evidence. Evolutionary biologists are honing in on the microevolutionary processes that affect phenotypes as new technologies for studying alleles and developmental pathways, for example, become available. Mutation and natural selection have been observed in experimentally controlled populations of Drosophila and in wild populations, from weeds, such as Palmer amaranth, to oldfield mice. Advances in the study of biogeography and paleontology, on the other hand, have been instrumental to the understanding of macroevolutionary processes. Speciation has been observed in apple maggot flies, Rhagoletis pomonella, for example, and ring species provide just one example of the processes that can lead to evolution at this level. In addition, the plethora of fossil evidence that is constantly being discovered, including a diversity of transitional forms, is providing evidence for the originations and extinctions of species of the historic past.

    The evidence for the theory of evolution comes from a myriad of different sources, and all the various lines of evidence serve to support that theory. Scientists may disagree on specific aspects of both microevolution (e.g., the role of allopatry and sympatry in speciation) and macroevolution (e.g., the causes of the mass extinction at the end of the Cretaceous) and how microevolution and macroevolution interact, but neither microevolution nor macroevolution has more or less evidentiary support.

2. What lines of evidence have macroevolutionary biologists used to determine the origin of marsupials? How are these lines distinct?

    Macroevolutionary biologists used three very different lines of evidence to determine the origin of marsupials. The first line of evidence came from an understanding of microevolutionary processes. Scientists started by developing a molecular phylogeny of the living marsupials. They found that marsupials form a monophyletic clade, indicating that all living marsupials are more related to each other than they are to other mammals—marsupials share a single common ancestor. This marsupial clade is also completely nested within a clade of mammals found only in South America, so the common ancestor of marsupials and their nearest relatives likely originated in the South American region.

The second line of evidence came from the fossil record. Fossil marsupials are known from Asia, Europe, North America, South America, and Antarctica—not just Australia. Scientists were able to construct a phylogeny of the major marsupial lineages using morphological traits of fossil marsupials. Just like the molecular phylogeny, Australian marsupials formed a monophyletic clade. But unlike the molecular phylogeny, the phylogeny based on fossils indicated that marsupials from Antarctica were more closely related to marsupials from Australia than marsupials from South America. Fossils from North American marsupials were the most distantly related. The ages of fossils indicate they originated in Asia, dispersed to North America, and then to Europe (and even northern Africa) and to South America, Antarctica, and Australia.

The third line of evidence came from geologists and the reconstruction of the Earth's continents through plate tectonics. Over a hundred million years ago, Asia and North America were linked as were South America, Antarctica, and Australia. Dispersal from Asia to North America and then to South America, Antarctica, and Australia would have been relatively unimpeded. These continents started to drift apart tens of millions of years ago, which would have isolated populations of marsupials in Antarctica and Australia.

These lines of evidence are distinctly different—molecular biology, paleontology, and geology—but together they provide evidence to support the phylogenetic hypothesis for the origination, diversification, and extinction of the marsupials.

3. Are punctuated equilibria and Darwin's theory of natural selection at odds in evolutionary theory?

No. Punctuated equilibria is a model of change over time. It proposes that although most lineages do not change much over the course of their geologic history, every once in a while there are periods of rapid change, often leading to speciation events. Darwin

proposed that gradual changes accumulate over extended periods of time, ultimately leading to speciation events.

Many of those that try to fault evolution in general, and macroevolution in particular, set up a false argument that punctuated equilibria and Darwin's idea of gradual change are alternative theories. They imply that both cannot be true. But just like macroevolution and microevolution, this argument focuses on the patterns—not the underlying mechanisms. Darwin proposed that organisms struggle for existence—those with traits that help them do better (that is, survive and/or reproduce) have more offspring, and the traits become more common in the population in future generations. Although he didn't have an understanding of mutations, genes, or genetic toolkits, he understood that traits were valuable commodities in the struggle. If a mutation, say to an important gene in the genetic toolkit, led to an altered development of fins, that trait may rapidly spread through a portion of the population if those that possess the new limb structure do well. (Maybe they can access a new food source that they couldn't get to without limbs they could put weight on and use to muck around at the water's edge.) Changes may continue to accumulate in the genetic structure of the limb relatively rapidly (remember, rapid change in the geologic record can still be hundreds of thousands of years), again, because individuals with incrementally better and better limbs do relatively better than the others in the population. That set of traits may work for a while—maybe with small gradual changes accumulating that are not as detectable (especially in the fossil record)—a long period of stasis. Then another structure-changing mutation may occur that allows some individuals to take in oxygen differently. And rapid change occurs again.

Also, the tree of life is not linear. It is not a simple path from fish to frogs to lizards to birds to mammals, and finally, to humans (see Figure 4.6). A trait that allows exploitation of a new habitat may lead to rapid adaptive radiations—radiations that can occur on small or large scales. But because of the fossil record, scientists may never find more than one or two representatives of any radiation. They have to piece together the tree of life from an entirely incomplete fossil record, but the more they look, the more they find. They are gradually filling in gaps in our understanding, and both models offer insight to deciphering the patterns of the history of life.

4. How is the term Cambrian Explosion misleading?

The term Cambrian Explosion is misleading because it implies that a diversity of animals sprang into existence all at once. First, the Cambrian Explosion did not occur on a single date 542 million years ago, nor did it occur within a single year, decade, century, or even millennia. The Cambrian Explosion can be thought of as a 23-million-year (plus or minus) event where animal taxa rapidly diversified (again, "rapid" is a relative term in geology). Second, animals did not suddenly pop into existence. Animals began appearing in the fossil record more than 635 million years ago—that's nearly 200 million years before the date humans attach to the Cambrian Explosion. In fact, estimates based on molecular clocks indicate that animals likely first appeared on Earth around 800 million years ago, with major groups splitting off 600 and 700 million years ago. Sometime around 542 million years ago, however, evolutionary processes led to the huge diversification of groups found in the fossil record. The molecular clock estimates predict that scientists should be able to find fossil animals in rocks up to or older than 800 million years old; macroevolution predicts that the Cambrian Explosion may be related to changing physical conditions and a new kind of ecosystem; and microevolution predicts that the Cambrian Explosion may be related to the evolution of a versatile genetic toolkit.

5. Should humans be concerned about the pace of extinctions of organisms that are not directly related to our survival?

Yes. Although a single extinction may not have significant effects on an ecosystem, mass extinctions can have cascading effects that may ultimately affect organisms that humans rely on directly. Indeed, indirect costs may be higher than we can even imagine. Even though humans may be able to control some of the organisms necessary for our survival through farming and ranching, if ecosystems begin collapsing at a similar pace to extinctions in Earth's past, humans will likely be affected. Whether that effect is dramatic population decline or extinction is unknown. However, whether we choose to excessively contribute to the pace of extinctions or not is under our control.

Test Yourself

1. a; 2. d; 3. b; 4. d; 5. d; 6. a; 7. a; 8. c; 9. e; 10. e

Chapter 12
INTIMATE PARTNERSHIPS:
HOW SPECIES ADAPT TO EACH OTHER

Check Your Understanding

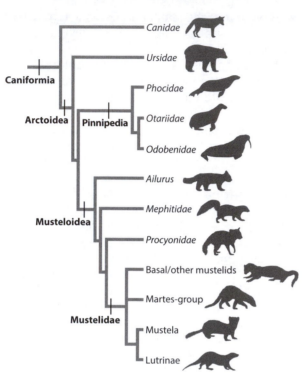

1. Which of the following statements are depicted by this phylogeny (from Flynn et al. 2005):
a. Otters (Lutrinae) evolved from martins (*Martes* group).

b. The ancestors of otters (Lutrinae) became gradually more "otter-like" over time.
c. Living otters (Lutrinae) represent the end of a lineage of animals whose common ancestor was wolf-like.
d. Otters (Lutrinae) share a common ancestor with Odobenidae (walruses).
e. All of the above are depicted by this phylogeny.

2. What is directional selection?
a. Selection that favors individuals at one end of the distribution of a trait.
b. Selection that favors individuals with a trait near the mean of the population.
c. Selection that guides the evolution of the traits organisms need.
d. Selection that leads to the distribution of species in a phylogeny.

3. Which of the following is NOT a factor necessary for evolution to occur by natural selection?
a. Variation among individuals.
b. Differential success.
c. Heredity.
d. Randomness.

Learning Objectives for Chapter 12:

➢ Define coevolution.
➢ Describe the geographic mosaic theory of evolution.
➢ Explain antagonistic coevolution and mutualism.
➢ Compare and contrast the coevolutionary arms race between snakes and newts with the coevolutionary arms race between rabbits and the myxoma virus.
➢ Explain diversifying coevolution.
➢ Describe the evolution of mitochondria.
➢ Explain how mobile genetic elements and retroviruses affect genomes.

Identify Key Terms

Connect the following terms with their definitions on the right:

antagonistic coevolution	A coevolutionary relationship where an increase in fitness of one species can potentially increase the fitness of a partner species.
coevolution	A type of mimicry that occurs when several harmful/distasteful species resemble each other in appearance, facilitating the learned avoidance by predators.
coevolutionary arms race (escalation)	Occurs when species interact antagonistically in a way that results in each species exerting directional selection on the other. As one species evolves to overcome the weapons of the other, it, in turn, selects for new weaponry in its opponent.
diffuse coevolution	A theory that proposes that the geographic structure of populations is central to the dynamics of coevolution. The structure and intensity of coevolution vary from population to population, and coevolved genes from these populations mix together as a result of gene flow.
endosymbionts	Sausage-shaped structures within eukaryotic cells that produce energy from oxygen, sugar, and other molecules.
geographic mosaic theory of coevolution	Selection that occurs in two species as a result of their interactions with each other.
mitochondria	Mutualist organisms that live within the body or cells of another organism.
Müllerian mimicry	A coevolutionary relationship where an increase in fitness of one species can potentially decrease the fitness of a partner species.
mutualism	Reciprocal evolutionary change between interacting species, driven by natural selection.
reciprocal selection	A broad term that often refers to mutualistic relationships, but may include other coevolutionary relationships, such as parasitism and commensalism.
retrovirus	A condition where many species interact with many other species leading to coevolution.
symbiosis	An RNA virus that becomes part of the host cells' DNA, for example, the virus that causes AIDS, the human immunodeficiency virus (HIV).

The Tangled Bank Study Guide

Link Concepts

1. Fill in the following concept map with the key terms from the chapter:

mutalism

antagonism

geographic mosaic theory

coevolution

retrovirus

coevolutionary arms race

endosymbiont

diffuse coevolution

mitochondria

Müllerian mimicry

mobile genetic element

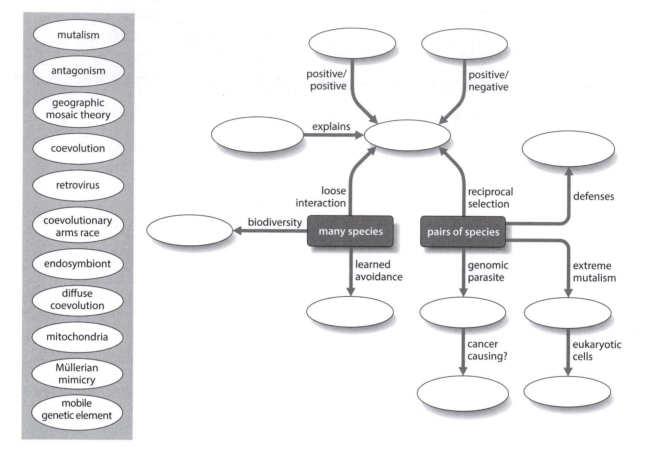

positive/positive   positive/negative

explains

loose interaction   reciprocal selection   defenses

biodiversity   many species   pairs of species

learned avoidance   genomic parasite   extreme mutalism

cancer causing?   eukaryotic cells

2. Develop your own concept map that explains the geographic mosaic theory of coevolution. Think about hotspots, coldspots, gene flow, variation among individuals, heredity, differential survival/reproduction, reciprocal selection, extinction, and speciation.

Interpret the Data

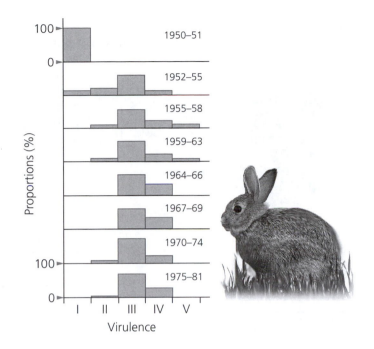

Figure 12.8 shows the changes in virulence in myxoma virus over time. Myxoma virus was introduced into Australia to control the exploding rabbit population. Virulence is measured as "virulence grades": I (deadliest) to V (mildest). Each bar represents a proportion (0–100 percent) of the total virus population during the specific time period (the scales have been left off to make the graphs easier to read).

What was the most common virulence grade in 1951? In 1952?

Explain how the virulence grades could change so quickly between 1951 and 1952.

How stable is the intermediate grade of virulence in the myxoma virus?

Games and Exercises

Plants and Their Pollinators

Pollination is incredibly important for sexual reproduction in plants (see Chapter 9 to learn how important sexual reproduction is in evolution). Pollination is simply the transfer of pollen grains from the stamen of one flower to the stigma of the same or another flower. Coevolution of plants and their pollinators has resulted in an amazing diversity of flowers and pollinators, each with adaptations that favor traits in the other. You can examine these adaptations at work simply by going outside and making careful observations of pollination in progress.

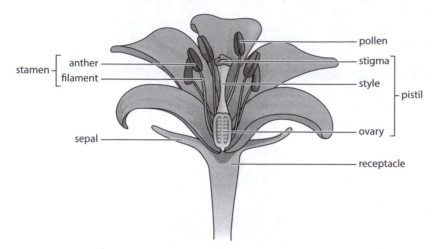

Category	# Observed
Bees	
Butterflies	
Moths	
Beetles	
Flies	
Carrion-eating flies	
Ants	

Start by finding a patch of similar looking flowers—a square meter would be a standard size you could observe easily and replicate elsewhere. Spend 10 minutes observing pollinators on the flowers. Record your observations in a table like this one:

Use the Key to Identifying Common Pollinators to help identify some of these major groups. Go through the key before observing insects so you can become familiar with important characteristics necessary to identify groups. To read the key, start with 1a; if the characteristic fits, follow the directions in the right-hand column. If they don't fit, go to 1b. (The insect should fit either 1a or 1b—do not go to 2 unless one of those characters fits.) If you have time, repeat your observations with another patch of flowers that have a different form.

Key to Identifying Common Pollinators

1a.	Wings not visible, or hard wing covers concealing wings	go to 2
1b.	Wings visible	go to 3
2a.	No wings, narrow area between thorax and abdomen	Ants
2b.	Hard wing covers conceal flight wings, form line down middle of back, chewing mouthparts	Beetles
3a.	One set of filamentous wings, eyes large and obvious (careful, some Syrphid flies mimic bees)	Flies
3b.	Two sets of wings	go to 4
4a.	Both sets of wings often colorful, covered with scales	go to 5
4b.	Wings membranous, usually clear	go to 6
5a.	Antennae with knob-like ends, wings usually folded when at rest	Butterflies
5b.	Antennae with feathered ends, no knob, wings often open at rest	Moths
6a.	Thorax and abdomen joined by narrow "waist," abdomen often pointed	Wasps

6b.	"Waist" not as marked, body usually hairy	Bees
7a.	Pollen carried on "belly"	Megachilid leafcutting bee, orchard mason bee
7b.	Pollen carried mainly on leg	go to 8
8a.	Usually small (~5–10 mm), black or metallic green, short tongued	Halictids "sweat bees," Andrenids
8b.	Long tongue, usually over 12 mm	go to 9
9a.	Spur on hind leg, abdomen often appears striped	Anthrophorid digger bees
9b.	No spur, body robust, usually over 20 mm, yellow and black, eyes not hairy	Bumblebees *Bombus*
9c.	No spur, golden brown color, 12–15 mm, hairy, eyeballs	Honeybees *Apis mellifera*

- Describe the flowers you observed (think about shape, size, and color).
- What time of day did you make the observation?
- What was the weather like when you were making your observations?
- Was one group of insects more common on your patch than other groups? Which?

Pollinators are not *trying* to pollinate the flower. As they collect pollen or nectar for food, they brush up against a flower's pistil accidentally and transfer some of the pollen that has collected on their bodies. "Pollination syndromes" are basically the suites of floral traits that attract certain pollinators. Traits include shape, color, size, floral arrangement, and such. Different flowers tend to attract different insect pollinators. Bees tend to prefer yellow, blue, and purple flowers; there are hundreds of types of bees that come in a variety of sizes and have a range of flower preferences. Butterflies tend to prefer red, orange, yellow, pink, blue flowers. They need to land before feeding, so they like flat-topped clusters (e.g., zinnias, calendulas, butterfly weeds) in a sunny location. Moths, on the other hand, pollinate white or dull-colored fragrant flowers since they can't see colors (e.g., potatoes, roses). Flies tend to prefer simple bowl-shaped flowers or clusters in green, white, and cream, and carrion-eating flies pollinate maroon and brown flowers with foul odors (e.g., wild ginger). Although ants like pollen and nectar, they aren't good pollinators, and many flowers have sticky hairs or other mechanisms that keep them out of nectar receptacles.

Some other pollinators to consider are hummingbirds and bats. Hummingbirds use red, orange, purple/red tubular flowers with lots of nectar, since they live exclusively on flowers, but they don't need a landing pad because they hover while feeding. They commonly pollinate sages, fuchsias, honeysuckles, nasturtiums, columbines, jewelweeds, and bee balms. Bats are nocturnal, and they don't have the greatest eyesight. They do have a keen sense of smell, however, so flowers pollinated by bats must be large, light-colored, and night-blooming with a strong fruity odor (e.g., many cactus flowers).

You can also use a dichotomous key to determine the likely pollinators of flowers of different shape, size, and color:

Pollination Syndromes Dichotomous Key

1a.	Flowers small, inconspicuous and usually green or dull in color, petals reduced or absent	Wind
1b.	Flowers conspicuous, usually with white or colored petals	go to 2
2a.	Flowers regular in shape, radially symmetrical	go to 3
2b.	Flowers irregular in shape, bilaterally symmetrical	go to 9
3a.	Flowers purple brown or greenish in color, often with strong odor of rotting fruit or meat, little floral depth	go to 4
3b.	Flowers with little odor, or sweet odor	go to 5
4a.	Flowers purple brown, sometimes with a "light window"	Flies
4b.	Odor day or night, dull color	Beetles

5a.	Flowers with deep corolla tube	go to 6
5b.	Flowers more dish shaped, reward accessible, yellow, or with abundant pollen	Bees, flies, small moths
6a.	Flowers red, open in day, little or no odor, no nectar guide, nectar plentiful	Hummingbirds
6b.	Flowers not pure red, usually sweet odor	go to 7
7a.	Flowers yellow, blue, or purple, corolla tube not narrow, but sometimes needing forced opening, often with nectar guides	Long-tongued bees
7b.	Flowers red, purple, or white, corolla tube or spur narrow, usually lack nectar guide	go to 8
8a.	Flowers purple or pink, diurnal, upright, with landing area	Butterflies
8b.	Flowers white or pale, pendant, open or producing odor at night	Moths
9a.	Flowers red, little or no odor	Hummingbirds
9b.	Flowers with odor, usually with nectar guides	Bees

- What are some of the characteristics of flowers that may affect access to pollen and nectar?

- Do the characteristics of pollinators affect the survival and reproduction of flowers?

- What happens if a pollinator lands on the "wrong" plant?

- Are all pollinators limited to certain kinds of plants? Why or why not?

Adapted from Pollination Ecology: Field Studies of Insect Visitation and Pollen Transfer Rates by Judy Parrish (http://tiee.ecoed.net/vol/v2/experiments /pollinate/pdf/pollinate.pdf) and "A Plethora of Pollinators" by Alison Perkins (www.bioed.org/ ECOS/inquiries/inquiries/PlethoraofPollinators.pdf).

Go Online

Orchid Bees—Euglossa

Orchids and their pollinators are an incredible example of the extreme adaptations that can evolve in two intimately linked species. This video from Hila Science Video illustrates the remarkable adaptations in both orchids and bees that evolved and the fitness benefits that drove that evolution. Are these "perfect" adaptations though?

http://www.youtube.com/watch?v=gEcv3dBuOe4

Evolution: Toxic Newts

This clip from PBS *Evolution*, *"Evolutionary Arms Race,"* produced by WGBH, takes you out in the field with the Brodie's as they study the evolutionary arms race between newts and garter snakes. The interaction of these two species is driving their evolution; the toxin produced by a single newt can kill 12 people, and although snakes are resistant, that resistance comes with a cost.

http://www.pbs.org/wgbh/evolution/library/01/3/l_013_07.html

Evolution: Ancient Farmers of the Amazon

Another clip from PBS *Evolution*, *"Evolutionary Arms Race,"* produced by WGBH, illustrates a classic example of coevolution of species based on a mutualistic relationship. Cameron R. Currie explains that the relationship between ants and the fungus they "farm" is not so simple; an evolutionary arms race is shaping the species' interactions as well.

http://www.pbs.org/wgbh/evolution/library/01/3/l_013_01.html

 Goby-Shrimp Mutualism

This short video shows a unique mutualistic relationship between two marine species and the research that helped scientists understand this relationship (see the "About" tab under the video).

https://sites.google.com/site/coralreefsystems/videos/short-movies/synb

---

 Don't Eat Me! I'm with Those Guys

In 2001, scientists found convincing evidence for the survival advantage of Müllerian mimics.

http://news.sciencemag.org/2001/01/dont-eat-me-im-those-guys

Endosymbiosis

In this YouTube video, Paul Andersen explains the evidence for symbiosis and the evolution of eukaryotes.

http://www.youtube.com/watch?v=-FQmAnmLZtE

Common Misconceptions

Coevolution and Convergent Evolution

Convergent evolution occurs when similar traits evolve in unrelated lineages. Dolphins and fish both have streamlined bodies for swimming, fins, and fanned tails. These organisms converged on similar forms even though they are not closely related. Coevolution is the reciprocal evolutionary change between interacting species, driven by natural selection. When dolphins eat fish, both predators and prey can coevolve. But the important difference to think about is pattern versus process. Convergent evolution describes the similarities in the traits of organisms, and examining these traits can help scientists understand the genetic and developmental underpinnings of similar adaptations. Coevolution explains why interacting species change.

Coevolution Does Not Create Mutual Harmony in Nature

Evolution and coevolution cannot create harmony in nature—harmony is a human construct. As Darwin noted, life is a struggle for existence. Mutualistic relationships occur when an increase in fitness of one species can potentially increase the fitness of a partner species. If circumstances were to change, however, such as the arrival of a new species, there is no evolutionary guarantee that the mutualism(s) that originally evolved will be sustained. In addition, coevolution can result from antagonistic relationships—predators eating prey, parasites eating individuals from the inside out, nest parasites pushing less developed offspring out of their parent's nest.

Coevolution Does Not Promote Stable Coexistence of Species

Although scientists have shown that coevolution can make the dynamics within broad communities of species more stable, coevolution does not promote the stable coexistence of species. Coevolution requires that partner species be able to respond to changes in each other—one must be exerting selective pressure on the other. The random nature of mutations, however, along with constraints imposed by developmental biology, does not guarantee any response. Costs or constraints may prevent proportional responses, and stable coexistence does not necessarily follow. For example, a new mutation in an antagonistic relationship may lead to such high fitness in one partner that the other cannot survive or reproduce at all. Snakes and newts may currently be in an evolutionary arms race, but a change in a single amino acid can make snakes resistant to newt toxin. Newts may not be able to respond to that new selective pressure because making deadlier toxins requires a change in many pathways. Alternatively, coevolution may actually speed up rates of evolutionary responses, promoting biodiversity and reducing the likelihood of stable coexistence. For example, the evolutionary arms race between milkweed toxins and the caterpillars that eat them may have caused both milkweed plants and butterflies (adult caterpillars) to diversify.

Another important factor to consider is that coevolution can intensify extinctions and the loss of biological diversity. Extinctions can be caused by a variety of abiotic and biotic factors. Depending on the coevolutionary relationships among species, the loss of one species may lead to the loss of many other species as well.

Contemplate

Could you use the principles of geographic mosaic theory to examine how the uniforms of sports teams change over time? How? (Think about cities with multiple sports teams—especially the same sport—versus other cities.)	Do horticulturalists (people who cultivate plants for human use) affect coevolutionary partnerships in the natural world? Why or why not?

Delve Deeper

1. How do variation among individuals, differential survival or reproduction, and heredity act to generate the patterns of newt toxicity and snake resistance observed by the Brodies?

2. Why is antagonistic coevolution important to the development of evolutionary arms races?

3. How does the interaction between long-tongued flies and *Zaluzianskya* flowers fit into geographic mosaic theory?

4. How can diversifying coevolution lead to speciation?

5. Are scientists absolutely certain about the origin of the relationship between eukaryotes and mitochondria?

Go the Distance: Examine the Primary Literature

Scott Carroll and his colleagues were able to show that evolutionary responses can be rapid. They examined the evolution of beak size in soapberry bugs as this native species encountered an invasive exotic species, the balloon vine.

- What evidence did Carroll and his colleagues use to examine changes in soapberry bug beak size?

- What proportion of the population were they able to evaluate?

- How often could they evaluate changes in beak size?

Carroll, S. P., J. E. Loye, H. Dingle, M. Mathieson, T. R. Famula, and M. P. Zalucki. 2005. And the Beak Shall Inherit—Evolution in Response to Invasion. *Ecology Letters* 8 (9): 944–51. doi:10.1111/j.1461-0248.2005.00800.x. http://onlinelibrary.wiley.com/doi/10.1111/j.1461-0248.2005.00800.x/abstract.

Test Yourself

1. What is coevolution?
   a. When one species needs to change because another species that it depended on changed.
   b. When two or more species interact, and each acts as an agent of selection causing evolution of the other.
   c. Evolutionary changes that are shared among two or more species, such as wings.
   d. Evolutionary changes among two or more species that are complementary, so that all species benefit.
   e. All of the above.

2. What are the key conditions necessary for the change in beak lengths in soapberry bugs across Australia?
   a. Greater food resources provided by large balloon vine fruits, opportunity to feed on balloon vines, and ability to reproduce as a result of better food.
   b. Variation among individuals that is heritable and leads to differential survival or reproduction.
   c. Need for a food resource, and long periods of time.
   d. A length of association between soapberry bugs and balloon vines similar to their native fruits.

3. Which of these statements about the virulence of rabbit myxoma virus is TRUE?
   a. Rabbit myxoma virus needed to become less virulent so it could coexist with its rabbit hosts.
   b. Directional selection favored a coevolutionary escalation where resistance evolved in rabbits and less virulence evolved in the virus.
   c. Rabbit myxoma virus became less virulent over time because natural selection favored strains that did not immediately kill the rabbit hosts, enhancing the likelihood of spreading and infecting other rabbits.

   d. None of the above is a true statement.
   e. All of the above are true statements.

4. Müllerian mimicry can be considered:
   a. Convergent evolution.
   b. Coevolution.
   c. Speciation.
   d. Both a and b.
   e. None of the above.

5. What does the geographic mosaic theory predict about coevolutionary outcomes?
   a. Because reciprocal selection can vary across geographic space and time, coevolutionary outcomes can also vary across geographic space and time.
   b. Coevolution will only occur in hotspots.
   c. Organisms are less likely to coevolve because gene flow will reduce the effects of fitness outcomes that result from even strong reciprocal selection.
   d. Reciprocal selection has to be strong across the entire range of both interacting species for coevolution of the species to occur.
   e. Hotspots and coldspots will vary in the amount of gene flow that separates them.

6. What did Sandra Anderson and her colleagues demonstrate by manually pollinating the flowers of the native *Rhabdothamnus solandri* on mainland North Island New Zealand?
   a. That mainland flowers could produce more fruit than island flowers.
   b. That mainland flowers were not producing as much fruit as they were capable of producing.
   c. That within island populations, *Rhabdothamnus solandri* flowers were producing fruit at near maximum levels.
   d. That the diversity of pollinators was higher on the mainland.
   e. All of the above.

7. Rapid extinction of a group of species can result when
   a. Species are antagonistic coevolutionary relationships.
   b. Alleles are swept to fixation.
   c. Species are in mutualistic coevolutionary relationships.
   d. Selection favors highly virulent strains of a virus.
   e. None of the above.

8. What evidence did Nancy Moran use to support her hypothesis about the evolution of the endosymbionts inside the glassy-winged sharpshooter?
   a. She compared phylogenies of the endosymbionts and their insect hosts and found the patterns of lineages were closely matched.
   b. She used fossils of sap-feeding insects and sharpshooters with known ages to establish time frames for the endosymbiotic relationships.
   c. She compared the molecular phylogenies of two endosymbionts, *Sulcia* and *Baumannia*, and found similar patterns in their lineages.
   d. She used all of this evidence to support her hypothesis.

9. What is a retrovirus?
   a. A virus that is fairly old.
   b. An RNA virus that uses an enzyme to become part of the host cells' DNA.
   c. A virus that has been resurrected from extinction.
   d. A virus that only replicates using RNA.

10. Why do you think different people carry slightly different versions of the same retrovirus segment in their DNA?
    a. Because after the virus invaded the human common ancestor, different versions evolved in different lineages of humans.
    b. Because the genome for every human cell is different.
    c. Because there are almost 100,000 fragments of endogenous retroviral DNA in the human genome, and scientist can't tease them apart very easily.
    d. Because retroviruses became mobile genetic elements in the human genome that can insert themselves anywhere in the genome.
    e. Because after the virus invaded the human common ancestor, variants that could replicate in somatic cells fared better than those that could replicate only in germ-line cells.

Answers for Chapter 12

Check Your Understanding

1. Which of the following statements are depicted by this phylogeny:

   a. Otters (Lutrinae) evolved from martins (*Martes* group).

   Incorrect. Phylogenies are often incorrectly interpreted as a ladder of evolution—one species evolving into the next. However, think of the phylogeny as fluid, with the parts at every intersection swinging independently. The Lutrinae/Procyonidae (raccoons, ringtails, and the like) branch could easily flip, so that it might appear that otters evolved "from" skunks (Mephitidae). So the correct way to interpret the relationships is to follow the lineage back to the node—those groups share a common ancestor (Chapter 4).

   b. The ancestors of otters (Lutrinae) became gradually more "otter-like" over time.

   Incorrect. Phylogenies represent the relationships among populations, genes, or species. The clades are organized so that the taxa within them share derived characteristics. So otters share more derived characteristics with the Mephitidae (skunks) than they do with Canidae (wolves). That doesn't mean that organisms were becoming more "otter-like" and less "wolf-like." The visualization of the phylogeny is fluid, with the parts at every intersection swinging independently. Any of the branches from Ursidae (bears) to Lutrinae can easily be flipped within the phylogeny, which eliminates that interpretation (Chapter 4).

   c. Living otters (Lutrinae) represent the end of a lineage of animals whose common ancestor was wolf-like.

   Incorrect. Phylogenies represent the relationships among populations, genes, or species. Each branch represents a lineage. Lineages are organized in clades, so that the taxa within them share derived characteristics. Otters (Lutrinae) share more derived characteristics with walruses (Odobenidae) than they do with wolves (Canidae). But otters do not represent the end of the Caniformia lineage, and the common ancestor was not necessarily wolf-like. Many lineages of Caniformia have gone extinct, and many still exist. The trick is figuring out how all these lineages are related (Chapter 4).

   d. Otters (Lutrinae) share a common ancestor with Odobenidae (walruses).

   Correct. Phylogenies represent the relationships among populations, genes, or species. Each branch represents a lineage. Lineages are organized in clades, so that the taxa within them share derived characteristics. The visualization of the phylogeny is fluid, with the parts at every intersection swinging independently. Any of the branches from Ursidae (bears) to Lutrinae can easily be flipped within the phylogeny. The correct way to interpret the relationships is to follow the lineage back to the node—those groups share a common ancestor (Chapter 4).

   e. All of the above are depicted by this phylogeny.

   Incorrect. Phylogenies are often incorrectly interpreted as a ladder of evolution or the end of a lineage because they can appear very ladder-like. However, think of the phylogeny as fluid, with the parts at every intersection swinging independently. Any of the branches from Ursidae (bears) to Lutrinae can easily be flipped within the phylogeny. Phylogenies represent the relationships among populations, genes, or species. Each branch represents a lineage. Lineages are organized in clades, so that the taxa within them share derived characteristics. The correct way to interpret the relationships is to follow the lineage back to the node—those groups share a common ancestor. Lineages below that node share derived characteristics (Chapter 4).

2. What is directional selection?

   a. Selection that favors individuals at one end of the distribution of a trait.

   Correct. Individuals within a population often have traits that vary in size, shape, color, or other characteristics. Usually these traits have a distribution around a mean (or average), so some individuals have smaller or larger traits (or shorter or longer, or greener or bluer) than other individuals. Directional selection is a type of natural selection that results when individuals at one end of the distribution have greater survival or reproductive success than other individuals. For example, perhaps individuals with smaller hind legs can swim faster and escape predators more easily than individuals with larger legs.

b. Selection that favors individuals with a trait near the mean of the population.

Incorrect. This form of fitness is known as stabilizing selection. Individuals within a population often have traits that vary in size, shape, color, or other characteristics. Usually these traits have a distribution around a mean (or average), so some individuals have smaller or larger traits (or shorter or longer, or greener or bluer) than other individuals. When individuals in the center of the distribution have greater survival or reproductive success than other individuals, selection is *stabilizing*. For example, birds with bills that are too small or too large may not be able to eat as many types of foods as individuals with intermediate sized bills. Directional selection, on the other hand, is a type of natural selection that results when individuals at one end of the distribution have greater survival or reproductive success than other individuals.

c. Selection that guides the evolution of the traits organisms need.

Incorrect. Natural selection and fitness cannot predict an organism's needs. Individuals have traits, and those traits can vary among individuals—some longer or shorter, some bigger or smaller, some greener or bluer than others. The measurement of a specific trait for all individuals in the population has a distribution, with a mean (or average) and spreading out from there. Directional selection is a type of natural selection that results when individuals at one end of the distribution have greater survival or reproductive success than other individuals. For example, perhaps individuals with smaller hind legs can swim faster and escape predators more easily than individuals with larger legs.

d. Selection that leads to the distribution of species in a phylogeny.

Incorrect. Directional selection is a form of fitness that favors individuals at one end of the distribution of a trait. Individuals within a population often have traits that vary in size, shape, color, or other characteristics. Usually these traits have a distribution around a mean (or average), so some individuals have smaller

or larger traits (or shorter or longer, or greener or bluer) than other individuals. Different species and even different populations can experience different forms of directional selection, but the distribution of species in a phylogeny does not reflect a single type of selection. Phylogenies are hypotheses about the historical relationships among species— common ancestors and their descendants. A variety of microevolutionary and macroevolutionary processes, including directional selection, disruptive selection, genetic drift, and extinction, ultimately shapes the diversity of species depicted by a phylogeny.

3. Which of the following is <u>NOT</u> a factor necessary for evolution to occur by natural selection?

a. Variation among individuals.

Incorrect. The variation among individuals in some trait value is the raw material on which natural selection can act. When individuals vary, some may do better or worse than others as a result of those trait values. Those that do better will leave relatively more offspring, and more of their alleles will be represented in the next generation—thus evolution—or change over time—by natural selection.

b. Differential success.

Incorrect. Individuals that differ in their survival or reproductive success ultimately leave different numbers of offspring in future generations. When individuals vary in some trait value, some individuals may do better or worse than others as a result of those trait values. Those that do better will leave relatively more offspring, and more of their alleles will be represented in the next generation—thus evolution—or change over time—by natural selection.

c. Heredity.

Incorrect. Individuals may vary in some trait value, and some individuals may do better than others as a result of those traits. As a result, those better individuals may leave more offspring than poorer individuals. However, if those trait values do not have some component that is heritable, the trait cannot be passed on, allele frequencies cannot change, and evolution by natural selection cannot occur.

d. Randomness.

Correct. Evolution by natural selection is *not* random. Natural selection acts on the heritable variation among individuals. Mutations, however, are random. Any time a cell replicates its DNA, there is a small chance that a random error will occur. Mutations are the ultimate source of heritable genetic variation. However, although the input of new variation by mutation is random, the subsequent action of selection on that variation is anything but random.

Identify Key Terms

antagonistic coevolution	A coevolutionary relationship where an increase in fitness of one species can potentially decrease the fitness of a partner species.
coevolution	Reciprocal evolutionary change between interacting species, driven by natural selection.
coevolutionary arms race (escalation)	Occurs when species interact antagonistically in a way that results in each species exerting directional selection on the other. As one species evolves to overcome the weapons of the other, it, in turn, selects for new weaponry in its opponent.
diffuse coevolution	A condition where many species interact with many other species leading to coevolution.
endosymbionts	Mutualist organisms that live within the body or cells of another organism.
geographic mosaic theory of coevolution	A theory that proposes that the geographic structure of populations is central to the dynamics of coevolution. The structure and intensity of coevolution vary from population to population, and coevolved genes from these populations mix together as a result of gene flow.
mitochondria	Sausage-shaped structures within eukaryotic cells that produce energy from oxygen, sugar, and other molecules.
Müllerian mimicry	A type of mimicry that occurs when several harmful/distasteful species resemble each other in appearance, facilitating the learned avoidance by predators.
mutualism	A coevolutionary relationship where an increase in fitness of one species can potentially increase the fitness of a partner species.
reciprocal selection	Selection that occurs in two species as a result of their interactions with one another.
retrovirus	An RNA virus that becomes part of the host cells' DNA, for example, the virus that causes AIDS, the human immunodeficiency virus (HIV).
symbiosis	A broad term that often refers to mutualistic relationships but may include other coevolutionary relationships, such as parasitism and commensalism.

Link Concepts

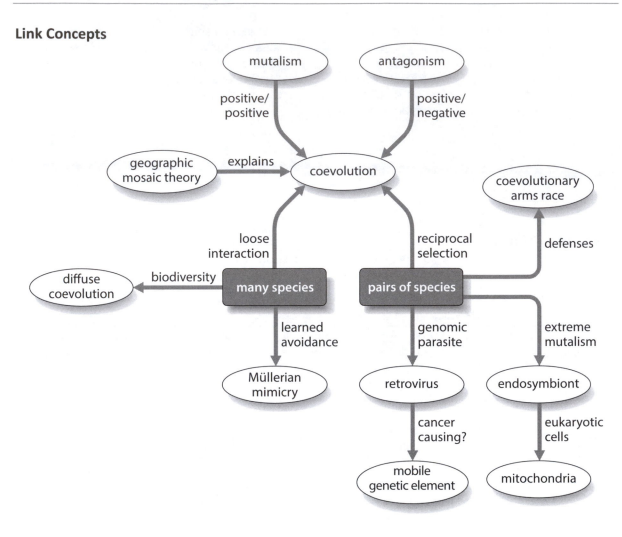

Interpret the Data

- What was the most common virulence grade in 1951? In 1952?
  Grade I. Grade III.
- Explain how the virulence grades could change so quickly between 1951 and 1952?
  Myxoma virus is not a single individual, nor a single strain. Viruses can evolve quickly because of high reproductive rates—mutations can spread through a virus population producing new strains rather quickly. Any particular rabbit is likely to be infected by more than one strain of myxoma virus, and competition among strains within that rabbit would lead to increased virulence. However, individual rabbits infected with strains that become too virulent may die before the virus can be transmitted—any virus population within a dead individual is a dead strain. So strains that aren't too virulent will be more successful because their populations can grow and still be transmitted to new hosts.
      When the virus was first introduced, rabbits were abundant, and virulence was high because selection favored competition among strains. As rabbit populations plummeted, hosts became scarce and selection for increased transmission became dominant. Those strains that did not kill their hosts immediately (grades II–V) did better and became more common in the rabbit population.
- How stable is the intermediate grade of virulence in the myxoma virus?
  Fairly stable, but not absolute. Selection clearly favors the intermediate grades, but that doesn't mean something couldn't alter the balance. Virus populations are still experiencing mutations,

and one of those could lead to entirely unknown trajectories. In fact, the data show that the balance does indeed shift—a more virulent grade appeared for a short time from 1970 to 1974, although it was only a small proportion of the total virus population.

Delve Deeper

1. How do variation among individuals, differential survival or reproduction, and heredity act to generate the patterns of newt toxicity and snake resistance observed by the Brodies?

   Variation among individuals, differential survival or reproduction, and heredity are all necessary for the coevolution of newts and snakes. Within populations of newts, some individuals may be more toxic than other individuals, and within snake populations, some individuals may be more resistant to that toxin than others. In some areas where the various individuals of each species interact, some newts may be less likely to be eaten by snakes (and therefore produce more offspring) than other newts, and some snakes may be able to eat more (and therefore produce more offspring) than other snakes. But because resistance affects a snake's ability to move, snakes that are highly resistant to the toxin may be more likely to be eaten by their predators (and therefore produce fewer offspring) than other snakes. So differential survival and reproduction affect the relative number of offspring individuals produce in the next generation. Heredity is the means by which the traits that influence survival and reproduction are transmitted to the next generation. Offspring of toxic newts are likely to be toxic, and offspring of resistant snakes are likely to be resistant.

   More importantly, toxic newts influence the survival and reproduction of snakes, and resistant snakes influence the survival and reproduction of newts. The interaction of the two species affects the frequencies of the alleles for each of these traits in the next generation. Hotspots occur where toxicity and resistance are evolving rapidly because this kind of natural selection is strong. Coldspots occur where the two species are not coevolving (perhaps because predation on slow, resistant snakes is high enough to overcome the survival and reproductive benefits of being able to eat newts).

2. Why is antagonistic coevolution important to the development of evolutionary arms races?

   Antagonistic coevolution is important to the development of evolutionary arms races because if the interacting organisms did not affect each other, then neither would be affected by changes in the other. Individuals would not experience better or worse reproduction or survival as a result of changes in individuals of other populations, and the frequency of alleles that they carried would not be affected as a result. Evolutionary arms races require conflict—the changes in one of the interacting populations of organisms must affect the fitness of the other. Once a mutation arises and spreads through a population in response to the selection exerted by the other, however, for the coevolutionary interaction to continue, that other population must experience a beneficial mutation that arises and spreads in response to selection exerted by the first. Evolutionary arms races result because each population acts on the other, and each population responds to the selective pressure.

3. How does the interaction between long-tongued flies and *Zaluzianskya* flowers fit into geographic mosaic theory?

   Long-tongued flies and *Zaluzianskya* flowers have a mutualistic relationship; as the flies drink nectar from the flowers, they pick up pollen on their heads, which they transfer to other flowers as they continue to feed. The flies' tongues and the flower tube depth match at different sites—they are long where long-tubed *Zaluzianskya* flowers are common, and short where short-tubed *Zaluzianskya* flowers are common—so there's a tremendous range of variation in both the length of long-tongued flies and the depth of *Zaluzianskya* flower tubes. The type of selection and the strength of reciprocal selection depend on whether *Zaluzianskya* flowers are the dominant source of nectar for the flies or not, however. Where *Zaluzianskya* are common, directional selection coupled with strong reciprocal selection leads to long tongues and long tubes. But where other flowers are common, competition for pollinators dilutes the strength of reciprocal selection, and directional selection favors shorter tubes and shorter tongues. A geographic mosaic results, where strong mutualistic relationships are common in some locations and not in others.

4. How can diversifying coevolution lead to speciation?

Diversifying coevolution is the increase in genetic diversity caused by the heterogeneity of coevolutionary processes across the range of ecological partners. Speciation is one predicted outcome of geographic mosaic theory. When local populations differ in the details of their coevolutionary partnerships, or in the strength or direction of selection resulting from those coevolutionary interactions, they may diverge in ways that reduce the likelihood of exchange of gametes between them. These reproductive barriers to gene flow will further reduce the likelihood of interbreeding between individuals of different populations. Depending on the strength and types of selection, one evolutionary lineage may become two or more lineages, representing a speciation event. Coevolution, because it can drive unusually rapid evolution, may dramatically accelerate the speed with which such divergence arises, and thus may accelerate the rate of speciation.

5. Are scientists absolutely certain about the origin of the relationship between eukaryotes and mitochondria?

No. Scientists are not absolutely certain about anything. Scientists rely on the best available evidence to support or refute hypotheses. For example, the phylogenies of both eukaryotes and bacteria are hypotheses about relationships based on the best evidence currently available. As more tools to examine genomes become available and scientists continue to unravel the components of genomes, their understanding of these relationships may change. Alternatively, the evidence may continue to support the proposed relationships, leading to greater and greater confidence.

More importantly, science is not about *proving* anything. DNA evidence, for example, indicates that mitochondrial DNA is not like any eukaryote's DNA, human or any other animal. Mitochondrial DNA is more similar to bacterial DNA, and specifically a lineage of bacteria found in marine environments. Using evidence from Giardia and other eukaryotes, scientists propose that ancestors of these marine bacteria invaded ancestral eukaryotes, affecting the survival and reproductive success of each, and ultimately leading to the lineages of all modern eukaryotes.

Giardia is a great example of how new evidence altered our current understanding. Until scientists could identify mitosomes in *Giardia* as remnants of mitochondria, some had proposed that the relationship came after the evolution of single-celled protozoans, not before as is currently accepted. But this evidence only changed how scientists looked at the timing of this coevolutionary relationship. It didn't change our understanding of coevolutionary relationships or phylogenetic relationships or evolution. In fact, taken together, all these different lines of evidence provide strong support for the hypothesis that mitochondria evolved from an early coevolutionary relationship between bacteria and eukaryotes.

Test Yourself:

1. b; 2. b; 3. c; 4. d; 5. a; 6. b; 7. c; 8. d; 9. b; 10. a

Chapter 13
MINDS AND MICROBES: THE EVOLUTION OF BEHAVIOR

Check Your Understanding

1. Which of the following might be considered homologous traits?
   a. The bone structure in a human hand and a bat's wing.
   b. The gene called *Bmp4* in mice and the same gene in flies called Dpp.
   c. C-opsin (ciliary opsins) in the eyes of sharks and the eyes of hummingbirds.
   d. All of the above.
   e. a and c only.
   f. None of the above.

2. Why is an understanding of the genetic toolkit important to understanding the evolution of traits?
   a. Because the same underlying networks of genes govern the development of all animals, and new traits evolve as a result of mutations to genes within that network.
   b. Because the genetic toolkit consists of all of the genes scientists have identified so far.
   c. Because gene networks within the toolkit act like "modules" that can be deployed in new developmental contexts, yielding novel traits.
   d. All of the above.
   e. a and c only.
   f. None of the above.

3. Which of the following is an example of how a mutation can be both beneficial and costly to an individual?
   a. A mutation that confers resistance to pesticides but also makes insects more vulnerable to predators.
   b. A mutation that causes developmental genes to code for greater or fewer than seven cervical vertebrae but reduces fitness.
   c. A mutation that increases the size of offspring but reduces the number of offspring an individual can have.
   d. All of the above.
   e. None of the above.

Learning Objectives for Chapter 13:

➤ Explain how a behavior, such as tameness, can evolve.

➤ Describe behaviors in two organisms that do not have brains.

➤ Use the homologous proteins found in sponges to explain the evolution of the human neural network.

➤ Explain why the brains of all organisms do not process the exact same information.

➤ Compare the evolution of an innate behavior with the evolution of learning.

➤ Explain why selection primarily operates at the level of individuals (rather than groups).

➤ List costs and benefits of living in a group.

➤ Demonstrate how inclusive fitness can lead to kin selection.

➤ Explain how cheating affects kin recognition in social groups.

➤ Review current hypotheses for the evolution of tool use.

Identify Key Terms

Connect the following terms with their definitions on the right:

altruism	Differential performance (fitness) of groups of individuals causes some groups to outcompete and replace other groups.
behavioral ecology	Selection arising from the indirect fitness benefits of helping relatives.
cerebral cortex	Traits where an allele produces three things: a recognizable phenotypic trait, the ability to recognize this trait in other individuals, and preferential treatment of individuals with the trait.
green beard effect	Behavior of one individual toward another that benefits that other individual at a cost to its own fitness.
group selection	A cell that transmits electrical and chemical signals from one of its ends to the other.
homology	The outer layer of the cerebrum of the mammalian brain. This layer occupies 90 percent of the human brain.
inclusive fitness	A light-sensing pigment.
individual selection	The science that explores the relationship between behavior, ecology, and evolution, to elucidate the adaptive significance of animal actions.
innate behavior	The similarity of characteristics in different species because of a common ancestor.
kin recognition	Differential performance (fitness) of individuals causes some genotypes to outcompete and replace other genotypes.
kin selection	A special junction between neurons where one neuron transfers signals via chemical neurotransmitters to another neuron.
neuron	Behaviors that are expressed independent of experience (like swimming for fish—although not for humans).
neurotransmitter	The potential ability of an animal to distinguish between genetic relatives and nonkin.
phytochrome	An individual's total fitness, including its own reproduction as well as any increase in the reproduction of its relatives due specifically to its own actions.
synapse	Chemicals that transmit signals from one neuron to another across a synapse.

Link Concepts

1. Fill in the following concept map with the key terms from the chapter:

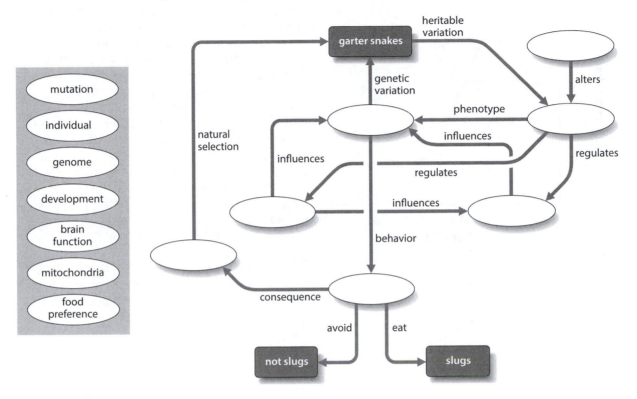

mutation

individual

genome

development

brain
function

mitochondria

food
preference

2. Develop your own concept map that explains the relationship between social behavior and kin
    selection. Think about kin recognition, cheaters, inclusive fitness, individual variation, heredity,
    green beards, and costs and benefits.

Interpret the Data

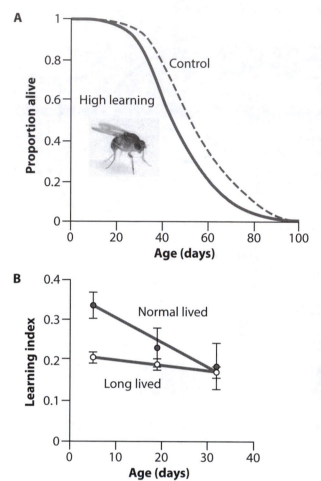

A

Proportion alive / Age (days)

Control

High learning

B

Learning index / Age (days)

Normal lived

Long lived

Tadeusz Kawecki and his colleagues at the University of Fribourg in Switzerland examined learning in fruit flies. They conducted an experiment in which they selected for flies that could learn quickly or were long lived, let these flies breed, and over the course of many generations, developed populations of high-learning flies and long-lived flies. A: When the scientists compared the high-learning populations to populations that had not been subjected to their artificial selection (the "controls"), they found that high-learning flies died sooner. B: When the scientists compared the flies selected for longevity, on the other hand, the long-lived flies turned out to be poor learners. (Adapted from Burger et al. 2008.)

What proportion of high-learning flies were alive at 50 days? What proportion of control flies were alive at 50 days?

What was the learning index for long-lived flies at 5 days? For normal flies?

Was the learning advantage of normal flies consistent over the lifetime of the flies?

Why did the scientists test the effects of learning with both a higher-learning experiment and a longer-lived experiment?

Games and Exercises

Making Observations

As humans, we are excellent observers. Our capacity for complex cognition allows us to quantify observations, discover patterns, and make predictions as we go through our daily life. Scientists do the same thing, but in a more formalized process. You can study behavior in animals simply by making careful observations within a formal framework.

Start with observations of a focal individual. Pick an organism that interests you. Make sure you can observe it continually for at least a short period of time. Note the date, time, location, weather, and habitat. Identify the individual if you can (a squirrel may have a stubby tail or a white patch) or by location. Record the sequence of behaviors, their duration, and any consequences (such as other individuals' responses). Record your data in a table like this one:

Individual ID	Behavior	Duration (minutes:seconds)

Are all the individuals behaving the same way? Do some spend more time behaving in one way than other ways? Can you tell the difference between males and females, and does gender affect behavior?

What Do Animals Learn?

If you have squirrels or crows in your neighborhood, they can be great tools for learning about learning. Find someplace they can be routinely found foraging or hanging out on the ground, like a park. Set up two experiments: one where you rush toward an individual, and one where you offer food. Think about time of day, presence of other individuals, presence of other people, weather, or any other factors that might influence their behavior. Be sure to use the same methods each time you conduct your experiment. When you rush at the squirrels or birds, do they fly away, move a little, run to a tree? How far away do they react? If you were to conduct the same experiment in the same place (and thus likely to sample the same squirrels, and maybe the same birds) over and over, do you think the reactions would change with more and more experience? What about if you approach slowly and offer the animals food (bird seed, for example)?

Here's what your data might look like:

Date	Sample	Reaction	Distance to reaction
Day 1	Focal Individual 1	Ran to tree	30 yards
Day 1	Focal Individual 2	Ran to tree	27 yards
Day 1	Focal Individual 3	Ran 20 yards	40 yards
Day 2	Focal Individual 1	Ran to tree	37 yards
Day 2	Focal Individual 2	Ran to tree	35 yards
Day 2	Focal Individual 3	Ran to tree	50 yards

Crows can recognize human faces. Scientists were able to discover this amazing feat with a simple experiment with surprising results. Check out the effects of masks on crow behavior:
http://www.livescience.com/14819-crows-learn-dangerous-faces.html.

Go Online

Video: Swimming Apes Caught on Tape

Swimming seems instinctive to some mammals, but not to the uninitiated human—or ape. But humans and apes can learn to swim, and this video shows that an orangutan named Suryia has some serious skill.

http://news.sciencemag.org/evolution/2013/08/video-swimming-apes-caught-tape

Slimy, But Not Stupid

Slime molds—eww. Ok, so not so much disgusting as interesting—they can actually navigate through a maze and figure out the shortest path to get to their food. Not bad for not having a brain.

http://news.sciencemag.org/2000/09/slimy-not-stupid

Dolphin Memories Span at Least 20 Years

Dolphins use whistles to communicate—they have their own unique whistles, and they can remember and repeat the whistles of other dolphins, apparently for 20 years or more!

http://news.sciencemag.org/plants-animals/2013/08/dolphin-memories-span-least-20-years

Kin Recognition in an Annual Plant

Kin recognition is well known in the animal kingdom, but plants behave, too. The amount of roots a plant grows should be a predictor of its competitive ability, at least below ground. This research shows that plants differ in the amount of roots they grow depending on whether they are planted with kin or not.

http://rsbl.royalsocietypublishing.org/content/3/4/435.full

Animal Behavior—Crash Course Biology #25

Hank Green explains animal behavior, including how and why behaviors evolve.

http://www.youtube.com/watch?v=EyyDq19Mi3A

Animal Behavior

 Paul Andersen explains the range of animal behaviors, from innate to learned, and the science behind understanding different types of behavior.

http://www.youtube.com/watch?v=6hREwakXmAo

Think Tank

The Smithsonian National Zoological Park developed this online exhibit about how to think about thinking. The site explores tools, language, and society with macaques, hermit crabs, orangutans, mangabeys, and rats.

http://nationalzoo.si.edu/Animals/ThinkTank/default.cfm

Frans de Waal: Moral Behavior in Animals

Frans de Waal shows that morality is not restricted to humans. Using video of research on animal behavior, he shows that two of the pillars of morality, reciprocity and empathy, can be found in many animals, from apes to elephants.

http://www.ted.com/talks/frans_de_waal_do_animals_have_morals.html

---

Common Misconceptions

What Is "for the Good of the Species"?

As humans, we cannot completely remove ourselves from the wiring in our brains. We have a cognitive capacity that allows us to make judgments about what is "good" and what is "bad"—that is our behavioral legacy. For example, we want to believe that selection has favored altruistic behavior for the "good" of the species. Evolution cannot recognize what is best for a group of organisms, however. Indeed, altruistic behavior is beneficial to the individual—it is "selfish." Individuals may benefit from altruistic behavior (directly or indirectly), and more of their alleles will be represented in the next generation as a result. Similarly, whether an allele spreads or disappears from a population is influenced more by how that allele affects an individual's fitness than how that same allele affects the fitness of the group. Only in rare circumstances can natural selection favor traits that are more beneficial to the group than to the individuals that carry them. Most of the time, these situations are unstable because any mutation that increases the fitness of an individual will spread quickly through the population. The immediate fitness consequences determine the success or failure of an allele, regardless of the ultimate outcome of the process.

Learning and the Capacity to Learn

Although humans, and other animals, clearly transmit learned behaviors through something akin to "culture," the behaviors themselves are not transmitted genetically—the ability to learn is. Teasing those two ideas apart is difficult because we humans are so adept at learning. But just like Lamarckian evolution, pierced ears, tattoos, and knowing all the words to the Best Song Ever are acquired during our lifetimes. Our offspring will not have pierced ears when they are born, but they may like piercings more because we raise them within that culture. The capacity to understand, mimic, and prefer that culture has a genetic component, however, and that capacity can be inherited. So some offspring may be more musically inclined than others, and they may be able to memorize words more easily, depending on the suites of genes inherited from their parents. The capacity for complex cognition can, and has, evolved over time.

Morality and Evolution

Humans are animals, and we share behaviors with animals. Those shared behaviors provide amazing evidence for the evolution of our species. Immorality, selfishness, and cruelty are our concepts of "wrong" and "right"—concepts that can be contemplated, because over time, natural selection favored genes that gave humans the capacity to think in abstract terms. We can decide what is wrong or what is right because of our evolutionary history. But natural selection and evolution have no inherent morality. Natural selection favors behaviors that increase survival and reproduction, and ultimately the spread of genes from one generation to the next. Depending on the conditions, cooperation and altruism may be winning strategies.

Contemplate

What's your family like at the holidays? What mechanisms do you have for kin recognition? Can you observe altruism? How might inclusive fitness be influencing your behaviors? Your siblings'? Your parents'?	Why might humans have evolved to be social?

Delve Deeper

1. As a behavioral ecologist, would you expect variation among individuals in an innate behavior, like pecking at a red dot? Based on your understanding, why might natural selection lead to the evolution of an innate behavior?

2. Why is group selection unlikely or rare?

3. In which system would you predict finding stronger kin recognition behaviors in males: a role-reversed breeding system where males assume all of the parental care or a breeding system where male paternal care increases the survival of the young?

4. What's the difference between learning and the evolution of learning behavior?

5. Why is understanding costs and benefits important to understanding the evolution of tool use?

Go the Distance: Examine the Primary Literature

In the wild, primates and birds can be pretty creative with tools, showing high degrees of innovation. Apes, corvids, and parrots are particularly innovative, but why these groups are exceptional is unclear. Alex Taylor and colleagues set out to examine cognition in New Caledonian crows (a species of corvid)— specifically, what capacity these birds had to problem solve, and what kind of cognitive mechanism can account for that capacity.

- How did their experiment address problem solving in New Caledonian crows?

- Was this a proximate study of behavior or an ultimate study?

- What was their sample size?

Taylor, A. H., D. Elliffe, G. R. Hunt, and R. D. Gray. 2010. Complex Cognition and Behavioural Innovation in New Caledonian Crows. *Proceedings of the Royal Society Series B: Biological Sciences* 277 (1694): 2637–43. doi:10.1098/rspb.2010.0285.
http://rspb.royalsocietypublishing.org/content/277/1694/2637.abstract.

Test Yourself

1. If cheating can be such a successful behavioral strategy, how can slime molds be cooperative?
   a. Group selection.
   b. Kin recognition.
   c. Inclusive fitness.
   d. All of the above.
   e. b and c only.
   f. Slime molds aren't cooperative.

2. What evidence would you use to support the hypothesis that the beginnings of the human nervous system probably evolved early in the animal lineage?
   a. The neurotransmitters that modulate the activity of neurons are nearly identical among animals with a nervous system.
   b. The sister group to all other animals that diverged 800 million years ago, the sponges, lacks a nervous system.
   c. The genes coding for proteins that send signals across a synapse from one neuron to another can be found in plants.
   d. All of the above support that hypothesis.
   e. None of the above supports that hypothesis.

3. What did Stevan Arnold do in his experiment to demonstrate that behavior can evolve?
   a. He manipulated the food preferences of young garter snakes by starving them until they switched from fish to slugs.
   b. He showed that allopatric speciation had led to inland and coastal species of garter snakes now isolated by their food preferences.
   c. He showed that food preferences varied, that variation was heritable, and that natural selection had affected food preferences in different portions of the snake's range.
   d. He showed that garter snakes can change their food preferences if they need to eat slugs instead of fish.

4. Which of the following is an example of behavior in plants?
   a. The trap of a Venus flytrap snaps shut in less than a second.
   b. Tobacco plants that produce nicotine and shunt it to their leaves when neighboring sagebrush are attacked by herbivores.
   c. Plants that release pheromones to attract natural enemies of the herbivores attacking them.

   d. All are examples of behavior in plants.
   e. Plants do not behave; seeing behavior in plants is anthropomorphic.

5. How does group selection differ from individual selection in the effect on alleles within a population?
   a. Group selection favors alleles that the group and the individual need to survive and do well.
   b. Group selection favors cooperation and individual selection does not.
   c. Group selection favors alleles that contribute to the performance of the group, not necessarily the individual.
   d. Group selection does not affect the alleles within a population; individual selection does affect the frequency of alleles because some individuals leave more surviving offspring that are more likely to carry those successful alleles than others.

6. Which of the following is NOT a cost of living in a group?
   a. Competition for food.
   b. Probability of catching a disease.
   c. Probability of being singled out by a predator.
   d. Competition for mates.
   e. Probability of being spotted by a predator.

7. Which of the following examples of altruism has no obvious benefit to the helping individual?
   a. A daughter helping her mother.
   b. A friend helping a friend.
   c. A stranger helping someone out of harm's way.
   d. A team member helping another team member.
   e. None of the above.

8. According to Figure 13.20, what is the coefficient of relatedness between two first cousins?
   a. 0.05.
   b. 0.125.
   c. 0.25.
   d. 0.5.
   e. It depends on which alleles are shared.

9. Are all amniotes capable of the same kinds of complex cognition?
    a. Yes. All amniotes can learn, so natural selection should shape their brains similarly when complex cognition is beneficial.
    b. Yes. Because amniotes share a common ancestor, the structure of their brains is similar, so they should all be capable of similar kinds of cognition.
    c. No. All amniotes can learn, but natural selection will shape brains for complex cognition only when it is beneficial to the species.
    d. No. Even though amniotes share a common ancestor, different types of complex cognition have evolved independently in distinct amniotic lineages.

10. Why might complex social cognition be related to tool use?
    a. Because complex social cognition may lead to extended interaction between individuals and cultural learning about tools.
    b. Because animals with complex social cognition skills live in large groups where tool use is common.
    c. Because animals with complex social cognition have larger brains than animals that live in small social groups.
    d. Because individuals with more social connections are more likely to use tools.
    e. Because species that live in more socially connected groups are more likely to use tools.

Answers for Chapter 13
Check Your Understanding

1. Which of the following might be considered homologous traits?

   a. The bone structure in a human hand and a bat's wing.

   Correct, but so are other answers. The bone structure of the human hand and a bat's wing are homologous because the same set of bones can be found in each species. The humerus extends from the shoulder and joins at the elbow with the radius and the ulna. The same cluster of wrist bones can be found at the end of the radius and ulna, and the same five digits extend from the wrist. Over time, evolution and natural selection shaped the bones in the two species differently, so any given bone in one is somewhat different from the corresponding bone in other. Those differences don't change the fact that they are the same set of bones, however, and they correspond because humans and bats share a common ancestor (see Chapter 2).

   b. The gene called *Bmp4* in mice and the same gene in flies called *Dpp*.

   Correct, but so are other answers. *Bmp4* and *Dpp* are homologous because they are essentially the same version of a gene (they have a similar sequence and function), even though they are found in very different organisms (see Chapter 8). They are important patterning genes and are relatively conserved—deleterious mutations are quickly lost to natural selection—but that doesn't mean the gene sequences are identical. Homologous genes can still produce similar proteins (because different sets of codons can still code for the same amino acid, see Chapter 5). In fact, scientists have been able to use homologous genes from one organism to switch traits on and off in a completely different organism. The genes don't share the same name simply because they were discovered independently by the different researchers working on mice and flies.

   c. C-opsin (ciliary opsins) in the eyes of sharks and the eyes of hummingbirds.

   Correct, but so are other answers. C-opsins are very similar in all vertebrates, not just sharks and hummingbirds. They have the same basic molecular shape and are stored in a stack of disks, each of which grows out of a hairlike extension of the retina called a cilium.

   They also have the same function—they relay their signal from the stack of disks through a series of proteins called the phosphodiesterase pathway (see Chapter 8). These homologies not only provide clues as to the common ancestors of vertebrates, they also offer amazing clues to the evolution of the human eye.

   d. All of the above.

   Correct. All of these traits are homologous because even though they are found in different organisms, the traits share similar structure and function. The same set of bones can be found in both a human hand and a bat's wing. *Bmp4* in mice shares the same structure and function as *Dpp* in flies. C-opsins are remarkably similar in all vertebrates, and they function to capture light. These homologies reflect common descent— a shared ancestor with a similar trait. More importantly, they show how evolution is constrained by the limits of the traits it has to work with. New adaptations are modifications of what was already there, and often those modifications are not perfect. Human hands and human eyes are not flawless (we develop repetitive stress disorders, and we have blind spots), but they evolved because different forms led to different survival and reproductive success of the individuals possessing them (see Chapter 6).

   e. a and c only.

   Incorrect. All of these traits are homologous because even though they are found in different organisms, the traits share similar structure and function. *Bmp4* and *Dpp* are homologous because they are essentially the same version of a gene (they have a similar sequence and function), even though they are found in very different organisms (see Chapter 8). The genes don't share the same name simply because they were discovered independently by the different researchers working on mice and flies.

   f. None of the above.

   Incorrect. All of these traits are homologous because even though they are found in different organisms, the traits share similar structure and function. The same set of bones can be found in both a human hand and a bat's wing. *Bmp4* in mice shares the same structure and function as *Dpp* in flies. C-opsins are remarkably similar in all vertebrates, and they function to capture light. These homologies reflect common descent—a shared ancestor with a similar trait. More importantly, they show how evolution is constrained by the limits of the traits it has to work with. New adaptations are modifications of what was already there, and often those modifications are not perfect. Human hands and human eyes are not flawless (we develop repetitive stress disorders, and we have blind spots), but they evolved because different forms led to different survival and reproductive success of the individuals possessing them (see Chapter 6).

2. Why is an understanding of the genetic toolkit important to understanding the evolution of traits?

a. Because the same underlying networks of genes govern the development of all animals, and new traits evolve as a result of mutations to genes within that network.

Correct, but so are other answers. Patterning genes, such as *Hox* genes and limb-patterning networks, demarcate the geography of developing animals, determining the relative locations and sizes of body parts. For 570 million years, mutations, even those that lead to relatively subtle changes to the toolkit, have been able to generate tremendous diversity in the animal kingdom (see Chapter 8).

b. Because the genetic toolkit consists of all of the genes scientists have identified so far.

Incorrect. Scientists are just beginning to understand how genes in gene networks function to produce complex adaptations in the phenotype. The genetic toolkit refers to the complex of networks shared by an ancient common ancestor of all animals 570 million years ago (plants also appear to share a genetic toolkit). Patterning genes, such as *Hox* genes, function within networks that can be deployed in new developmental contexts as a result of mutation. For example, limb-patterning networks can be turned on or off leading to the development of long legs or no legs at all (see Chapter 8).

c. Because gene networks within the toolkit act like "modules" that can be deployed in new developmental contexts, yielding novel traits.

Correct, but so are other answers. Patterning genes, such as *Hox* genes, function within networks that can be deployed in new developmental contexts as a result of mutation, yielding novel traits. For example, limb-patterning networks can be turned on or off leading to the development of long legs or no legs at all (see Chapter 8).

d. All of the above.

Incorrect. The genetic toolkit is a subset of described genes. Specifically, it refers to the complex of networks shared by an ancient common ancestor of all animals 570 million years ago (plants also appear to share a genetic toolkit). Patterning genes, such as *Hox* genes and limb-patterning networks, demarcate the geography of developing animals, determining the relative locations and sizes of body parts. Within the toolkit, gene networks act as modules that can be deployed in new developmental contexts, yielding novel traits (see Chapter 8).

e. a and c only.

Correct. Patterning genes, such as *Hox* genes and limb-patterning networks, demarcate the geography of developing animals, determining the relative locations and sizes of body parts. Within the toolkit, gene networks act as modules that can be deployed in new developmental contexts, yielding novel traits. For 570 million years, mutations, even those that lead to relatively subtle changes to the toolkit, have been able to generate tremendous diversity in the animal kingdom (see Chapter 8).

f. None of the above.

Incorrect. Patterning genes, such as Hox genes and limb-patterning networks, demarcate the geography of developing animals, determining the relative locations and sizes of body parts. Within the toolkit, gene networks act as modules that can be deployed in new developmental contexts, yielding novel traits. For 570 million years, mutations, even those that lead to relatively subtle changes to the toolkit, have been able to generate tremendous diversity in the animal kingdom (see Chapter 8).

3. Which of the following is an example of how a mutation can be both beneficial and costly to an individual?

a. A mutation that confers resistance to pesticides but also makes insects more vulnerable to predators.

Correct, but so are other answers. Pesticides can be a powerful selective force, and a mutation that confers resistance can be very valuable to the fitness of an individual. But that mutation may come at a cost if the gene is pleiotropic (i.e., it affects the expression of many different phenotypic traits). For example, mosquitos that were more resistant to organophosphate insecticides were also more likely to be killed by predators (see Chapter 6).

b. A mutation that causes developmental genes to code for greater or fewer than seven cervical vertebrae but reduces fitness.

Correct, but so are other answers. Seven cervical vertebrae in the neck are pretty much the rule in vertebrates, whether the organism is a mouse or a giraffe. Only seven cervical vertebrae may be limiting to some animals, like giraffes whose long necks allow them access to food resources unavailable to other ungulates. A mutation that influenced the number of vertebrae could increase flexibility, improve access to food, and increase survival and reproduction. That kind of mutation to a developmental gene likely comes at a pretty big cost, however. For example, human offspring born with more than seven cervical vertebrae are much more likely to develop pediatric cancer, if they aren't stillborn (see Chapter 8). So a mutation that altered the number of cervical vertebrae in giraffes could be highly beneficial, but the fitness costs may actually be constraining evolution.

c. A mutation that increases the size of offspring but reduces the number of offspring an individual can have.

Correct, but so are other answers. Bigger offspring may be able to survive better than smaller offspring, so a mutation that causes an individual to have bigger offspring may be more successful in terms of fitness. On the other hand, if that mutation also reduces the number of young, that individual's fitness may be reduced. Life-history theory is all about the trade-offs that may occur as the result of mutations, and depending on the circumstances—high or low predation— natural selection may favor individuals with or without the mutation (see Chapter 9).

d. All of the above.

Correct. A mutation can be both beneficial and costly to an individual if it is pleiotropic (i.e., it affects the expression of many different phenotypic traits). Sometimes the mutation confers a fitness benefit that outweighs its cost (resistance to insecticides, Chapter 6), and natural selection favors the mutation in the population. Other times the costs completely outweigh any benefits (increased mortality, Chapter 8) or those benefits depend on the circumstances (bigger or more offspring, Chapter 9).

e. None of the above.

Incorrect. A mutation can be both beneficial and costly to an individual if it is pleiotropic (i.e., it affects the expression of many different phenotypic traits). Sometimes the mutation confers a fitness benefit that outweighs its cost (resistance to insecticides, Chapter 6), and natural selection favors the mutation in the population. Other times the costs completely outweigh any benefits (increased mortality, Chapter 8) or those benefits depend on the circumstances (bigger or more offspring, Chapter 9).

Identify Key Terms

altruism	Behavior of one individual toward another that benefits that other individual at a cost to its own fitness.
behavioral ecology	The science that explores the relationship between behavior, ecology, and evolution, to elucidate the adaptive significance of animal actions.
cerebral cortex	The outer layer of the cerebrum of the mammalian brain. This layer occupies 90 percent of the human brain.
green beard effect	Traits where an allele produces three things: a recognizable phenotypic trait, the ability to recognize this trait in other individuals, and preferential treatment of individuals with the trait.
group selection	Differential performance (fitness) of groups of individuals causes some groups to outcompete and replace other groups.
homology	The similarity of characteristics in different species because of a common ancestor.
inclusive fitness	An individual's total fitness, including its own reproduction as well as any increase in the reproduction of its relatives due specifically to its own actions.
individual selection	Differential performance (fitness) of individuals causes some genotypes to outcompete and replace other genotypes.
innate behavior	Behaviors that are expressed independent of experience (like swimming for fish—although not for humans).
kin recognition	The potential ability of an animal to distinguish between genetic relatives and nonkin.
kin selection	Selection arising from the indirect fitness benefits of helping relatives.
neuron	A cell that transmits electrical and chemical signals from one of its ends to the other.
neurotransmitter	Chemicals that transmit signals from one neuron to another across a synapse.
phytochrome	A light-sensing pigment.
synapse	A special junction between neurons where one neuron transfers signals via chemical neurotransmitters to another neuron.

Link Concepts

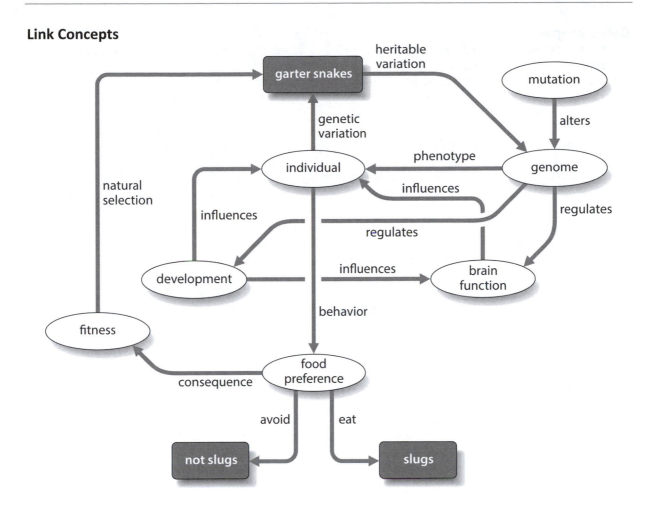

Interpret the Data

- What proportion of high-learning flies were alive at 50 days? What proportion of control flies were alive at 50 days?
  At 50 days, about 40 percent of high-learning flies, and about 60 percent of control flies were still alive. That's 20 percent fewer high-learning flies alive than control flies.
- What was the learning index for long-lived flies at 5 days? For normal flies?
  Approximately 0.21. Approximately 0.34.
- Was the learning advantage of normal flies consistent over the lifetime of the flies?
  No. Normal flies were better learners early in their lives, but that advantage disappeared as they aged. In fact, the learning index was not that different between normal flies and long-lived flies after flies reached 19 days old (notice that the error bars overlap, indicating that there was a lot of variation in the populations at that age).
- Why did the scientists test the effects of learning with both a higher-learning experiment and a longer-lived experiment?
  Because the results are correlations, the scientists could not determine that higher learning causes shorter life spans. By approaching the experiment from two directions—one that examines the effect of higher learning on life span and one that examines the effect of life span on learning—the scientists could compare the results. Each experiment supports the hypothesis that learning negatively affects life span, so the evidence is that much stronger.

Delve Deeper

1. As a behavioral ecologist, would you expect variation among individuals in an innate behavior, like pecking at a red dot? Based on your understanding, why might natural selection lead to the evolution of an innate behavior?

   Yes, even though a behavior is innate (i.e., a reliable behavioral response found in all members of a population), the behavior may be polygenic, resulting in a continuous phenotypic trait like running speed, rather than a discrete trait like eye color. So although all chicks peck at a red dot, some chicks might respond more quickly, some might respond more forcefully, and some might respond for a longer period of time than others. Or the behavior may be phenotypically plastic, varying in the response threshold to the contrast of colors. In either case, behavioral ecologists studying a particular innate behavior might look at the proximate mechanisms—the neurological or genetic framework—that is the underlying basis for the variation.

   The innate behavior may have an ultimate foundation, however, that led to its evolution. For example, pecking at a specific point on an adult's bill may elicit a feeding response—the sooner the chick can get fed, the sooner the adult can get back out hunting for more food, and the faster the chick will grow, leading to heavier offspring, better survival, and increased fitness. (Why that spot specifically would elicit feeding would be a proximate study.) Similarly, innate food preferences in inland garter snakes may have evolved to elicit rapid responses to quick-moving prey like fish and frogs. Responding quickly to a known food item could lead to greater foraging success and higher survival. (Understanding the neural architecture that causes garter snakes to respond to certain smells would be a study of the proximate mechanisms involved with prey capture. On the other hand, the phylogenetic relationship and how that may influence the development of the innate behavior would be an ultimate mechanism.)

2. Why is group selection unlikely or rare?

   Group selection is selection that favors alleles because they increase the performance of a group of individuals, causing groups with the allele to increase in size faster than groups that lack the allele. In principle, an allele could spread due to its group-benefiting effects even if it reduced the survival or reproductive success of individuals within the group bearing the allele. However, under most biologically relevant conditions, the fitness consequences to individuals will be much stronger than those arising from group selection, and alleles with group-benefiting effects generally only spread when they also act to increase the fitness of the individuals who bear them. Group selection is unlikely or rare because the sacrifice individuals may have to make to sustain the fitness of the group can easily be overwhelmed by natural selection on individuals. Any mutation that leads to an allele that causes an individual to behave selfishly will be more successful because that individual will experience greater survival or reproductive success than alleles that cause individuals to sacrifice themselves in terms of survival or reproductive success.

3. In which system would you predict finding stronger kin recognition behaviors in males: a role-reversed breeding system where males assume all of the parental care or a breeding system where male paternal care increases the survival of the young?

   In role-reversed mating systems where males assume all of the parental care, selective pressures to raise one's own offspring would be high. However, if behaviors in role-reversed systems, such as elaborate courtship with multiple bouts of copulation, increase the certainty of paternity, kin recognition behaviors would be redundant. As a result, the selective pressures for kin recognition as a means to raise a male's own offspring would be low.

   In breeding systems where male parental care increases offspring survival, the selective pressures to raise an individual's own offspring could also be high, especially if there are costs to the helping male (e.g., in terms of other opportunities to mate or even his own survival). Without the certainty of paternity that makes investing heavily in offspring cost effective, selection should heavily favor kin recognition behaviors. Therefore, behavioral ecologists would predict that male kin recognition would be stronger in these systems than in systems with alternative means for assuring paternity.

4. What's the difference between learning and the evolution of learning behavior?

Learning can be cultural—individuals can change their behavior based on observing other individuals. So information can be passed on from generation to generation, individuals can vary in what they've learned, and having learned can alter their behavior in ways that influence their fitness. But the learned behavior itself is, by definition, not directly inherited. But learning itself is not genetically controlled. Once you learn math, your offspring don't know math. You may have a better capacity to learn math, however, and that capacity can be transferred to some or all of your offspring—if it has a genetic component. The evolution of learning behavior is a population-level process. When learning behavior has a genetic component, that component can be passed on to future generations. If individuals vary in their capacity for complex cognition (say, to count the number of lions in a pack), and that ability confers some fitness advantage, evolution can shape learning behavior through natural selection.

5. Why is understanding costs and benefits important to understanding the evolution of tool use?

The evolution of tool use requires both proximate and ultimate mechanisms, each of which can be evaluated in terms of costs and benefits. For example, tool use requires recognition, goal setting, planning, and innovation. This complex cognition can be affected by context-specific patterns of gene expression, neural signaling, and physiology—all of which can come with costs that may affect expression of other genes, signaling pathways, or even the cell repair mechanisms leading to senescence. Research examining these types of proximate mechanisms has shown that there are indeed costs to the evolution of learning; fruit flies selected to learn quickly had shorter life spans than fruit flies that were not selected to learn quickly. Likewise, mammalian brain size relative to body size tends to be higher in species that are more social—where tool use may be more common. Large brains can be highly costly in terms of maintenance (e.g., brain temperature, oxygen regulation) and reproduction (e.g., live birth, parental care).

Understanding why tool use evolved is also related to the costs and benefits of that adaptation. Phylogenetic relationships provide insight to shared ancestry, but the evolution of tool use within lineages is not consistent. Not all animals with the faculties for complex cognition develop tools, and not all environments or ecological challenges may favor the evolution of tool use. For example, dogs and chimpanzees are both highly social species, but only dogs have the capacity to understand behaviors in other species (pointing by humans). Learning to use tools may evolve in unpredictable environments where a species cannot rely on innate responses. There are obvious costs to living in unpredictable environments (e.g., food supply), but the benefits may be less clear (e.g., enhanced nutrition). Similarly, the capacity for living in groups may lead to the ability to use tools, but sociality comes with its own costs and benefits. Natural selection favors learning and tool use only when the costs are outweighed by the benefits, and costs and benefits differ among species. So understanding costs and benefits may provide useful insights to the course evolution has taken in the development of these proximate and ultimate mechanisms.

Test Yourself

1. e; 2. d; 3. c; 4. d; 5. c; 6. e; 7. c; 8. b; 9. d; 10. a

Chapter 14
A NEW KIND OF APE

Check Your Understanding

1. Why are phylogenies such important tools in evolutionary biology?
   a. Because the relationships described by phylogenies are based on the best available evidence.
   b. Because the relationships described by phylogenies are based on different lines of evidence, including morphology, DNA, and fossils.
   c. Because the relationships described by phylogenies are hypothetical relationships that can be tested with additional evidence.
   d. Because the relationships described by phylogenies are developed with advanced statistical tools that can clarify complex relationships and generate additional hypotheses.
   e. All of the above.

2. Which of the following statements about molecular clocks is FALSE?
   a. Molecular clocks cannot be used to measure divergence in species separated by more than a few hundred million years.
   b. Molecular clocks must be calibrated because different types of DNA segments evolve at different rates.
   c. Molecular clocks result because within DNA, base-pair substitutions accumulate at a roughly clocklike rate, although substitution rates may differ between lineages.
   d. Molecular clocks can be used to predict and test the ages of unknown samples of DNA.
   e. None of the above is a false statement about molecular clocks.

3. Which is NOT a benefit of living in a social group?
   a. Reduced time spent watching for predators.
   b. Large brain size.
   c. Increased inclusive fitness.
   d. Smaller risk to individuals of being killed by a predator.
   e. Enhanced opportunities for learning.

Learning Objectives for Chapter 14:

➤ Explain the phylogenetic relationship between humans and chimpanzees.
➤ Describe the evidentiary support for the evolution of bipedalism.
➤ Compare and contrast tool use in chimpanzees and tool use in early humans.
➤ Compare and contrast the physiology of *Australopithecus sediba* and *Homo erectus*.
➤ Analyze the scientific debate about the placement of *Homo floresiensis* in the human lineage.
➤ Differentiate between *Homo heidelbergensis*, *Homo neanderthalensis*, and Homo sapiens.
➤ Explain how DNA can be used to examine the relationships among *Homo heidelbergensis*, *Homo neanderthalensis*, and *Homo sapiens*.
➤ Describe some of the selective pressures that led to the evolution of the human brain.

> ➤ Explain the evolution of *FOXP2*.
> ➤ Compare adaptations for living at high altitudes in Tibetans and Andean peoples.
> ➤ Discuss the evolutionary legacy of human emotions.
> ➤ Distinguish between the rational *Homo economicus* and the real *Homo sapiens* in terms of decision-making ability.
> ➤ Explain the role of MHC in mate choice.

Identify Key Terms

Connect the following terms with their definitions on the right:

Acheulean technology	Members of the "human" branch of the hominid clade, including the genus *Homo* and its close relatives, such as *Australopithecus*, but not chimpanzees and bonobos (*Pan*), gorillas (*Gorilla*), and orangutans (*Pongo*).
australopith	The co-option of a particular gene for a totally different function as a result of a mutation. The reorganization of a preexisting regulatory network can be a major evolutionary event.
bipedalism	A distinctive method for making large, teardrop-shaped stone tools called hand axes. The technique was much more sophisticated than earlier tool-making styles, but changed little from 1.7 million years ago to 100,000 years ago when hand axes disappeared from the fossil record.
gene duplication	A hypothesis that suggests that hominins, and humans in particular, evolved from other apes because selection favored ever-increasing social skills necessary for competing and cooperating in groups.
gene expression	A form of locomotion where an organism walks upright on two legs.
gene recruitment	A group of hominins that lived in Africa between 4 and 2 million years ago, including members of the genus *Australopithecus* (e.g., *Australopithecus afarensis*).
hominin	A Eurasian lineage of hominins that lived from 300,000 to about 28,000 years ago and likely interbred with humans.
Neanderthal	The process by which information from a gene is transformed into a product.
ultra-social hypothesis	A mutation that causes a segment of DNA to be copied a second time. The mutation can affect a region inside a gene, an entire gene, and in some cases, an entire genome.

Link Concepts

1. Map the Evidence

Literally hundreds of hominin fossils have been discovered since Darwin, some actually during his lifetime. New fossils are being discovered regularly. As a result, scientists are able to piece together a more and more detailed phylogeny for our species, but they still have questions. Some fossils are perplexing, either because not enough of the specimen was found, or because the bones don't fall neatly into our classification system. Some issues may never be resolved, but scientists are OK with that uncertainty. More importantly, the weight of all the fossil evidence, and evidence from other sources, is adding up—it's giving scientists a clearer picture of human evolution.

Whether issues exist about which fossil belongs to which species are completely resolved or not, you can use the fossil record as evidence for where hominins were and when they were there. Mapping out the locations will give you an understanding of how hominins diversified over time. The list of fossils below is a summary of actual fossils available for you to examine on the Smithsonian National Museum of Natural History's website on human origins (http://humanorigins.si.edu/). The list includes the fossil's approximate age, the site where the fossil was discovered, and the year of discovery. The list also includes the specimen name, so you can look up the fossil on the Human Origins website to examine images and descriptions.

Use the map following the table and some colored pens or pencils to track locations, ages, and species. For each species, pick a color and add a point to the center of the country where it was discovered. Make a notation about the age of the fossil.

age	species	site	year	specimen
7–6 million	*Sahelanthropus tchadensis*	Toros-Menalla, Chad	2001	TM 266-01-060-1
6 million	*Orrorin tugenensis*	Tugen Hills, Kenya	2001	BAR 1002'00
4.4 million	*Ardipithecus ramidus*	Aramis, Middle Awash, Ethiopia	1994	ARA-VP-6/500
4.1 million	*Australopithecus anamensis*	Kanapoi, Kenya	1994	KNM-KP 29285
3.5 million	*Australopithecus afarensis*	West Turkana, Kenya	1999	KNM-WT 40000
3.3 million	*Australopithecus afarensis*	Dikika, Ethiopia	2000	DIK-1-1
3.2 million	*Australopithecus afarensis*	Hadar, Ethiopia	1974	AL 288-1
3 million	*Australopithecus afarensis*	Hadar, Ethiopia	1992	AL 444-2
2.8 million	*Australopithecus africanus*	Taung, Republic of South Africa	1924	Taung Child
2.8–2.4 million	*Australopithecus africanus*	Sterkfontein, Republic of South Africa	1947	STS 71
2.8–2.4 million	*Australopithecus africanus*	Sterkfontein, South Africa	1971	STW 13
2.5 million	*Australopithecus africanus*	Sterkfontein, Republic of South Africa	1947	STS 14

2.5 million	*Australopithecus garhi*	Bouri, Middle Awash, Ethiopia	1999	BOU-VP-12/1
2.5 million	*Paranthropus aethiopicus*	West Turkana, Kenya	1985	KNM-WT 17000
2.5–2.1 million	*Australopithecus africanus*	Sterkfontein, Republic of South Africa	1947	STS 5
2.4 million	*Homo habilis*	Chemeron, Kenya	1967	KNM-BC 1
2.0–1.5 million	*Paranthropus robustus*	Drimolen, Republic of South Africa	1994	UW DNH 7
1.95 million	*Homo erectus*	Koobi Fora, Kenya	1984	KNM-ER 3228
1.95–1.78 million	*Australopithecus sediba*	Malapa	2008	MH1
1.95–1.78 million	*Australopithecus sediba*	Malapa	2008	MH2
1.9 million	*Homo habilis*	Koobi Fora, Kenya	1973	KNM-ER 1813
1.9 million	*Homo rudolfensis*	Koobi Fora, Kenya	1972	KNM-ER 1470
1.89 million	*Homo erectus*	Koobi Fora, Kenya	1972	KNM-ER 1481
1.8 million	*Homo erectus*	Koobi Fora, Kenya	1975	KNM-ER 3733
1.8 million	*Homo habilis*	Olduvai Gorge, Tanzania	1968	OH 24
1.8 million	*Homo habilis*	Olduvai Gorge, Tanzania	1960	OH 8
1.8 million	*Paranthropus boisei*	Olduvai Gorge, Tanzania	1959	OH 5
1.8–1.6 million	*Homo erectus*	Mojokerto, Java, Indonesia	1936	Mojokerto
1.8–1.5 million	*Homo habilis*	Swartkrans, Republic of South Africa	1969	SK 847
1.8–1.5 million	*Paranthropus robustus*	Swartkrans, Republic of South Africa	1936	SK 46
1.8–1.5 million	*Paranthropus robustus*	Swartkrans, Republic of South Africa	1950	SK 48
1.8–1.5 million	*Paranthropus robustus*	Swartkrans, Republic of South Africa		SK 54
1.77 million	*Homo erectus*	Dmanisi, Republic of Georgia	1999	D2282
1.77 million	*Homo erectus*	Dmanisi, Republic of Georgia		D3444
1.7 million	*Homo erectus*	Koobi Fora, Kenya	1974	KNM-ER 1808
1.7 million	*Homo habilis*	Koobi Fora, Kenya	1973	KNM-ER 1805
1.7 million	*Homo habilis*	Olduvai Gorge, Tanzania	1963	OH 16
1.7 million	*Paranthropus boisei*	Koobi Fora, Kenya	1969	KNM-ER 406
1.7 million	*Paranthropus boisei*	Koobi Fora, Kenya	1970	KNM-ER 732 A
1.6 million	*Homo erectus*	Koobi Fora, Kenya	1976	KNM-ER 3883
1.6 million	*Homo erectus*	Nariokotome, West Turkana, Kenya	1984	KNM-WT 15000
1.6 million	*Homo erectus*	Nariokotome, West Turkana, Kenya	1984	KNM-WT 15000
1.55 million	*Homo erectus*	Koobi Fora, Kenya	2000	KNM-ER 42700

1.4 million	Homo erectus	Olduvai Gorge, Tanzania	1960	OH 9
1.4 million	Paranthropus boisei	Konso, Ethiopia	1993	Konso KGA10-525
1.3–1.0 million	Homo erectus	Sangiran, Java, Indonesia	1969	Sangiran 17
> 1 million	Homo erectus	Sangiran, Java, Indonesia	1937	Sangiran 2
1 million	Homo erectus	Middle Awash, Ethiopia	1997	Daka BOU-VP-2/66
1 million	Homo erectus	Buia, Eritrea		Buia UA 31
1 million–700,000	Homo erectus	Trinil, Java, Indonesia	1891	Trinil 2
1 million–700,000	Homo heidelbergensis	Ceprano, Italy	1994	Ceprano
900,000	Homo erectus	Olorgesailie, Kenya	2003	KNM-OG 45500
780,000–400,000	Homo erectus	Zhoukoudian, China		Zhoukoudian
780,000–400,000	Homo erectus	Zhoukoudian, China		Zhoukoudian III
600,000	Homo heidelbergensis	Middle Awash, Ethiopia	1976	Bodo
500,000–200,000	Homo heidelbergensis	Elandsfontein, Republic of South Africa	1953	Saldanha
450,000	Homo heidelbergensis	Tautavel, France	1971	Arago 21
400,000–300,000	Homo erectus	Longtandau Cave, Anhui Province, China	1980	Hexian
350,000	Homo heidelbergensis	Steinheim, Germany	1933	Steinheim
350,000	Homo heidelbergensis	Lake Ndutu, Tanzania	1973	Ndutu
350,000–150,000	Homo heidelbergensis	Petralona, Greece		Petralona 1
300,000	Homo erectus	Narmada, India		Narmada
300,000–200,000	Homo heidelbergensis	West Turkana, Kenya		Eliye Springs ES11693
300,000–200,000	Homo neanderthalensis	Wadi Amud, Israel		Zuttiyeh
300,000–125,000	Homo heidelbergensis	Kabwe, Zambia	1921	Kabwe 1
259,000	Homo heidelbergensis	Florisbad, Republic of South Africa	1932	Florisbad
250,000–200,000	Homo erectus	Salé, Morocco	1971	Salé
250,000–70,000	Homo erectus	Solo River, Java, Indonesia		Ngandong 13
250,000–70,000	Homo erectus	Solo River, Java, Indonesia		Ngandong 14
250,000–70,000	Homo erectus	Solo River, Java, Indonesia		Ngandong 7
195,000	Homo sapiens	Omo River, Ethiopia	1967	Omo I
160,000	Homo sapiens	Jebel Irhoud, Morocco	1961	Irhoud 3
150,000–120,000	Homo sapiens	Singa, Sudan	1924	Singa
130,000	Homo heidelbergensis	Guangdong Province, China		Maba
130,000	Homo neanderthalensis	Krapina, Croatia	1899	Krapina 3

130,000–100,000	*Homo neanderthalensis*	Rome, Italy	1929	Saccopastore 1
122,000–50,000	*Homo neanderthalensis*	Mount Carmel, Israel		Tabun 1
120,000	*Homo sapiens*	Laetoli, Tanzania	1976	Ngaloba LH 18
120,000–80,000	*Homo sapiens*	Mount Carmel, Israel	1932	Skhul V
110,000–80,000	*Homo neanderthalensis*	Subalyuk Cave, Hungary		Subalyuk 2
100,000	*Homo sapiens*	Jebel Qafzeh, Israel	1933	Qafzeh 6
90,000–60,000	*Homo neanderthalensis*	Pech de l'Azé, France	1909	Pech de l'Azé I
70,000	*Homo neanderthalensis*	Bajsuntau, Uzbekistan	1938	Teshik-Tash
70,000–50,000	*Homo neanderthalensis*	Dederiyeh, Syria	1993	Dederiyeh 1
70,000–50,000	*Homo neanderthalensis*	La Ferrassie Cave, France	1909	La Ferrassie
70,000–45,000	*Homo neanderthalensis*	Forbes' Quarry, Gibraltar	1848	Gibraltar 1
68,000	*Homo sapiens*	Liujiang, China		Liujiang
65,000	*Homo neanderthalensis*	La Quina Rock Shelter, France	1915	La Quina 18
65,000	*Homo neanderthalensis*	La Quina Rock Shelter, France	1911	La Quina 5
60,000	*Homo neanderthalensis*	La Chapelle-aux-Saints, France	1908	La Chapelle-aux-Saints
45,000–39,000	*Homo sapiens*	Sarawak, Malaysia	1958	Niah Cave
45,000–35,000	*Homo neanderthalensis*	Shanidar, Iraq		Shanidar 1
41,500–39,500	*Homo sapiens*	Pestera cu Oase, Romania	2003	Oase 2
41,000	*Homo neanderthalensis*	Wadi Amud, Israel	1961	Amud
40,000	*Homo neanderthalensis*	Feldhofer Cave, Neander Valley, Germany	1856	Feldhofer
39,000–33,000	*Homo sapiens*	Hofmeyr, Republic of South Africa	1952	Hofmeyr
30,000	*Homo sapiens*	Cro-Magnon, France	1868	Cro-Magnon 1
36,000	*Homo neanderthalensis*	Saint-Césaire, France	1979	Saint-Césaire
18,000	*Homo floresiensis*	Liang Bua, Flores, Indonesia	2003	LB-1
13,000–9,000	*Homo sapiens*	Kow Swamp, Australia	1967	Kow Swamp
11,500	*Homo sapiens*	Minas Gerais, Brazil	1976	Lapa Vermelha IV Hominid 1
4,700	*Homo sapiens*	Tepexpan, Mexico	1947	Tepexpan 1
age uncertain	*Homo neanderthalensis*	Awirs Cave, Engis, Belgium	1829	Engis 2

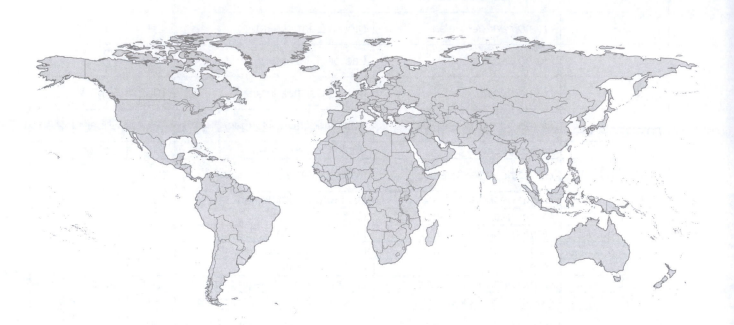

- What can you infer about the diversity of hominins from your map?

- Do you generally agree with the consensus of scientists below?

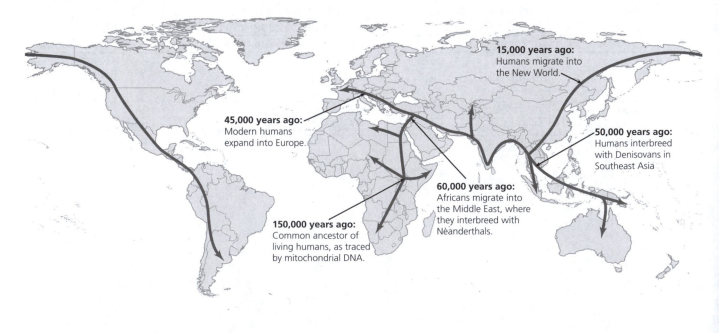

15,000 years ago: Humans migrate into the New World.

45,000 years ago: Modern humans expand into Europe.

50,000 years ago: Humans interbreed with Denisovans in Southeast Asia

60,000 years ago: Africans migrate into the Middle East, where they interbreed with Neanderthals.

150,000 years ago: Common ancestor of living humans, as traced by mitochondrial DNA.

2. Develop your own concept map that explains the genetic evidence for the evolution of humans. Think about complex cognition, Denisovans, gene duplication, genetic drift, humans, hybridization, language, mutations, Neanderthals, regulatory regions, sociality, and trichromatic vision.

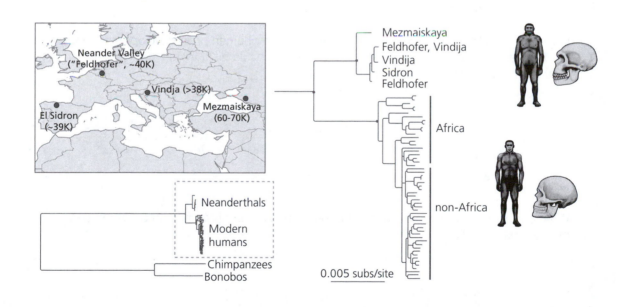

Interpret the Data

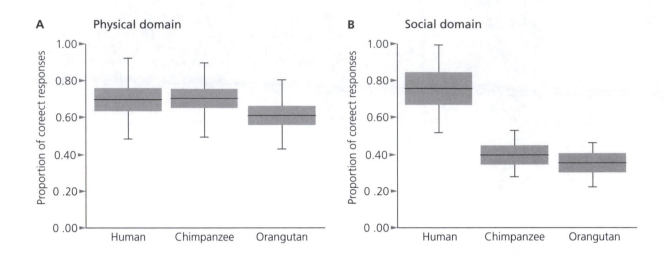

A Physical domain

B Social domain

Esther Herrmann and her colleagues (2007) tested the ultra-social hypothesis by examining the development of social skills in human children versus other apes. They performed two kinds of tests—one examining how children, chimpanzees, and orangutans responded to physical properties (such as space, quantity, and physical cause and effect), and the other examining how they responded to social cues (such as observation and learning).

If ultra-social relationships acted as selective agents on human brain development, would you predict that children, chimpanzees, and orangutans should differ in their understanding of basic math? Why or why not?

What do the width of the green bars and the "whiskers" for children, chimpanzees, and orangutans indicate?

If the whiskers indicate individuals with test results at the ends of the distributions of each species, how would you interpret the differences between children and chimpanzees in social skills test?

Games and Exercises

Tool making might not be unique to humans, but the depth and breadth of tools crafted by hominins are exceptional in the animal kingdom. Our hominin ancestors didn't have the mental capacity to craft highly specific tools. The evolution of stone tools reflects the evolution of the hominin brain. Stone technology changed over time as hominin lineages evolved the capacity to control their hands, to plan, and to create elaborate tools. Oldowan tools are very different from Acheulean technology. Our Homo sapiens ancestors were skilled craftspeople; their creations indicate a capacity for self-expression and trade.

Go online and search for images of stone tools. There is no shortage of images, and they are often grouped according to similarities (time, shape, function).

Bifacial chopper

Discoid

Polyhedron

0   5cm

Hammer stone

Flake scraper

Flake

Heavy-duty (core) scraper

Oldowan tools

Acheulean technology

- What characteristics distinguish older and newer technology?

- Can you guess what some of the tools might have been used for?

- Why might images of newer technology on the web be more prevalent than images of older technology?

- How might understanding the evolution of tool making apply to human's ability to fashion technology today?

Go Online

What Does It Mean to Be Human?

This exhibit from the Smithsonian Institution allows users to explore the research and the evidence scientists use to examine important questions about our evolutionary history.

http://humanorigins.si.edu/

Solving the Mystery of the Neandertals

Cold Springs Harbor Laboratory created a website to help students understand their own genetic origins (http://www.geneticorigins.org). You can participate in two different experiments using your own DNA. This animation explains the mitochondrial control region and shows how Svante Pääbo and his colleagues discovered the relationship between Neanderthals and humans.

http://www.geneticorigins.org/mito/media2.html

The Human Journey: Migration Routes

The Human Journey is an extensive project to map genetic markers in our population. Individuals can contribute their own DNA literally to become a part of the evidence. The resulting information can be used to trace our ancestry and understand our historical movements.

https://genographic.nationalgeographic.com/human-journey/

Becoming Human

This series explores what it means to be human. Each episode is one hour and available online from PBS *NOVA*.

http://www.pbs.org/wgbh/nova/evolution/becoming-human.html#becoming-human-part-1

http://www.pbs.org/wgbh/nova/evolution/becoming-human.html#becoming-human-part-2

http://www.pbs.org/wgbh/nova/evolution/becoming-human.html#becoming-human-part-3

Origins of Bipedalism

From PBS *NOVA*, this interactive runs through the diverse hypotheses for the evolution of bipedalism. After exploring each hypothesis, you can vote on the hypothesis that you think is the most likely.

http://www.pbs.org/wgbh/nova/evolution/origins-bipedalism.html

PBS *Evolution* and WGBH offer a wealth of videos and interactive websites related to human evolution. Here are a few links.

Finding Lucy

This video relives the excitement of the discovery of the fossil specimen known as Lucy.

http://www.pbs.org/wgbh/evolution/library/07/1/l_071_01.html

Humankind

This interactive explores the hominid family tree. The timeline shows details on species, details of their discovery, and time of existence.

http://www.pbs.org/wgbh/evolution/humans/humankind/

Walking Tall

Walking Tall is a short video that explains our evolutionary legacies—the adaptations that permit a graceful gait and the imperfections that lead to some our physical problems.

http://www.pbs.org/wgbh/evolution/library/07/1/l_071_02.html

Riddle of the Bones

 Another interactive explains the evidence scientists use to decipher fossil bones, including determining how ancient hominins moved, what they looked like, species relationships, and when they lived.

http://www.pbs.org/wgbh/evolution/humans/riddle/

Common Misconceptions

Humans Are Not Distinct "Kinds" of Organisms

Outside of science, people often want to separate species as clearly defined entities, but defining a species as a "distinct kind" does not provide any useful way of distinguishing groups of organisms, especially taxonomic levels. A "kind" could be fungi (a kingdom) or mammals (a class), a duck (an order) or a duck-billed platypus (a species). Defining exactly what constitutes a species is difficult, but scientists understand that classification is a human artifact. "Species" doesn't have to be defined precisely—unequivocally—for evolutionary theory to explain the existence of groups of organisms that share characteristics with a common ancestor.

Our species shares many characteristics with other hominins, and as scientists discover more and more fossil specimens, distinguishing each species is becoming more and more challenging. Our species *is unique* in that we, humans, are the only one of an apparently diverse group of hominins to survive. More importantly, however, scientists are uncovering the transitional fossils that together make up the human phylogenetic tree.

Humans Are Still Evolving

We tend to think our species has overcome the pressures of natural selection. It's true that we can modify our environments with technology, such as medicine, agriculture, and education. But natural selection remains powerful. Just as living at high altitudes and lactose tolerance shaped the evolution of mountain people and cattle herders in the not so distant past, humans will continue to face challenges to survival and reproduction. For example, women with a genetic tendency for low cholesterol are reproducing at greater rates in Framingham, Massachusetts, than women with a genetic tendency for high cholesterol. Predicting exactly what will shape us is nearly impossible, but the effects we are having on our climate and our environment may turn out to be strong selective factors in our future.

Phylogenetic Trees and "Advanced" Organisms

As humans, our need to classify also possessed us to categorize organisms as "primitive" and "advanced." Even when we look at phylogenetic trees, it's hard to escape believing that lineages that branch off earlier in the tree are "lower" or "primitive" forms. But branches can swing—like a mobile—and don't necessarily represent a "ladder-like" progression to more advanced species. As humans, we want to consider ourselves as the pinnacle of evolution. Our human brain is exceptional, and unique, but other adaptations in other organisms are exceptional as well. For example, the duck-billed platypus may seem "primitive," but many of its extraordinary traits (e.g., their unique bill) evolved after the lineage split off from other mammals (see Box 4.1).

Contemplate

Could you define different populations of humans today as different species? Why or why not?	How might our changing climate affect our evolution?

Delve Deeper

1. Which hypothesis has more support for the evolution of bipedalism: efficient feeding or heat dissipation?

2. Would you expect the common ancestor of chimpanzees and early hominins to have used tools? Why or why not?

3. What evidence indicates that our species, *Homo sapiens*, split from *Homo heidelbergensis* and not *Homo neanderthalensis*?

4. Which molecule(s) related to our emotions likely evolved early in our mammalian history?

5. Why isn't the human brain a rational decision-making organ like a computer?

Go the Distance: Examine the Primary Literature

Stephen Stearns and his colleagues tackle a question most of us think about when we contemplate human evolution: are humans still evolving?

- Besides the Framingham study, what are two other groups of individuals Stearns and his colleagues suggest looking at to measure fitness?

- What do those studies currently measure?

- According to Stearns and his colleagues, what are major unresolved issues facing scientists studying evolution in modern humans?

- Do Stearns and his colleagues contend that scientists can know how our species will evolve?

Stearns, S. C., S. G. Byars, D. R. Govindaraju, and D. Ewbank. 2010. Measuring Selection in Contemporary Human Populations. *Nature Reviews Genetics* 11 (9): 611–22. doi:10.1038/nrg2831.

Test Yourself

1. What features related to bipedalism do modern humans share with early hominin fossils?
   a. Similar anchors for muscles of the pelvis.
   b. Downward-pointing foramen magnum.
   c. Flattened toes.
   d. All of the above.
   e. None of the above.

2. Which of these phylogenies represent(s) the likely relationship between *Australopithecus sediba* and other hominins?

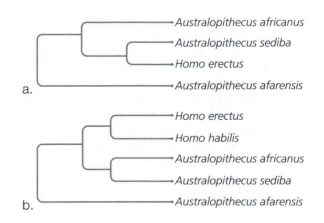

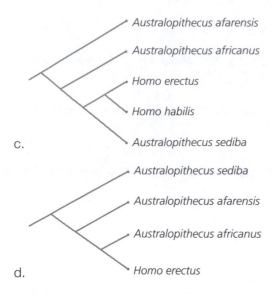

c.

d.

e. Both c and d represent the likely relationship.
f. Both a and c represent the likely relationship.

3. Why is the discovery of stone tools in the fossil record important in understanding the evolution of *Homo*?
   a. Because the new technology indicated a major transition in how hominin brains worked.
   b. Because stone tools added to the debate about when humans left Africa.
   c. Because the technology used to fashion stone tools also evolved.
   d. Both a and b.
   e. Both a and c.

4. What might additional fossils of early hominins add to our understanding of the origin of humans?
   a. Additional fossils could provide evidence to support the hypothesis that bipedal walking evolved before human ancestors left Africa.
   b. Additional fossils could provide evidence about the timing and speciation within the genus *Homo*.
   c. Additional fossils could provide evidence to support the hypothesis that *Homo erectus* shares a more recent common ancestor with *Homo floresiensis* and/or *Homo georgicus* than *Homo sapiens*.
   d. All of the above.
   e. None of the above.

5. Why do *Homo sapiens* and *Homo neanderthalensis* share DNA if not from common descent?
   a. Because *Homo sapiens* and *Homo neanderthalensis* hybridized during their coexistence.
   b. Because techniques for examining DNA are imperfect.
   c. Because *Homo sapiens* and *Homo neanderthalensis* are not really distinct species.
   d. Because of sampling error in the methodology.

6. Why does the imagery above lead to a misconception about the evolution of humans?
   a. Because it shows human evolution as a ladder-like progression, with one species transforming into another.
   b. Because it implies that we evolved from chimpanzees.
   c. Because it shows human evolution as a march of progress from primitive to advanced.
   d. Because it doesn't show the diversity of species that coexisted.
   e. All of the above.
   f. a and b only.

7. The hypothesis that contends that the evolution of large brain size in modern humans is a result of our ability to learn from others is called
   a. The cultural evolution hypothesis.
   b. The ultra-social hypothesis.
   c. The foramen magnum hypothesis.
   d. The social ape hypothesis.

8. Why isn't the human brain perfectly adapted to making modern-day decisions?
   a. Because components of the human nervous system evolved very early in our evolutionary history.
   b. Because complex adaptations develop from previously existing structures, pathways, and other traits.
   c. Because genetically based decision-making strategies that boosted our ancestors' chances of survival and reproduction were favored by natural selection over those that lowered fitness.
   d. Because rapid recent advances in technology have altered our environments within just a few dozen human generations, creating mismatches between our evolved decision-making strategies and the circumstances we now encounter.
   e. All of the above.

9. Is natural selection currently favoring lower cholesterol levels in women?
   a. No. Cholesterol is necessary for body maintenance and growth and generally causes heart disease only later in life; it is likely a result of senescence with very little effect on reproduction in women.
   b. No. Individual women vary in the tendency to metabolize cholesterol, but that variation is largely due to diet, and diet is more related to culture—not genetically transmitted.
   c. Yes. Women with greater body weight have greater reproductive success and lower cholesterol.
   d. Yes. Individual women vary in the tendency to metabolize cholesterol, this ability has a genetic component that can be transmitted to offspring, and individuals with a genetic tendency for low cholesterol have greater reproductive success than women with a genetic tendency for high cholesterol.

10. How does MHC function in mate choice in humans?
    a. Males whose MHC genetic diversity is dissimilar to a female's are more preferred than males whose MHC genetic diversity is similar.
    b. Women with high MHC tend to have sex for the first time at a younger age and to have more sexual partners.
    c. Males that are more physically attractive have less MHC genetic diversity and are preferred by females.
    d. Males that smell better to females have more MHC genetic diversity and are preferred by females.

11. How strong is the evidence supporting the hypothesis that multiple species of hominins existed at the same time?
    a. Very strong. Evolutionary biologists are confident in the ages of fossils, stone tools and other artifacts provide independent lines of evidence, and DNA analyses show evidence of hybridization.
    b. Strong, but scientists can't say for sure. Evolutionary biologists are finding new fossils regularly, and they can't know how that new evidence will affect current evidence.
    c. Weak. Evolutionary biologists don't have enough evidence about fossils or DNA to draw any conclusions.
    d. Inconclusive. The fossil record is incomplete, and the DNA evidence is based on a single bone.

Answers for Chapter 14
Check Your Understanding

1. Why are phylogenies such important tools in evolutionary biology?

   a. Because the relationships described by phylogenies are based on the best available evidence.

   Correct, but so are other answers. Phylogenies are developed using nested sets of shared derived characteristics based on the best available evidence at the time. However, scientists continue to discover new fossils, new species, and new ways to think about morphological and molecular evidence. They understand that phylogenies may shift and change, and they can use phylogenies to make predictions about future relationships (see Chapter 4).

   b. Because the relationships described by phylogenies are based on different lines of evidence, including morphology, DNA, and fossils.

   Correct, but so are other answers. Phylogenies can be developed using morphological, molecular, and even cultural evidence. Fossils (and other historical artifacts) can provide additional evidence for morphological relationships, DNA relationships, and the timing of branching events. Phylogenies are developed using nested sets of shared derived characteristics. The more independent lines of evidence used to build the relationships, the greater the likelihood that the relationships are accurate. Scientists understand that phylogenies may shift and change, however, and they can use phylogenies to make predictions about future relationships (see Chapter 4).

   c. Because the relationships described by phylogenies are hypothetical relationships that can be tested with additional evidence.

   Correct, but so are other answers. Phylogenies represent hypothetical historical relationships based on currently available evidence. They are explanations for patterns in nature that scientists can test with further evidence. Scientists understand that phylogenies may shift and change, however, and they can use phylogenies to make predictions about future relationships (see Chapter 4).

   d. Because the relationships described by phylogenies are developed with advanced statistical tools that can clarify complex relationships and generate additional hypotheses.

   Correct, but so are other answers. The information used to generate phylogenies can be voluminous, investigating a diversity of genes, populations, or species. Statistical methods help scientists resolve their data and develop the best possible hypotheses. These hypotheses can then be tested with additional evidence (see Chapter 7).

   e. All of the above.

   Correct. Phylogenies are developed using nested sets of shared derived characteristics based on the best available evidence at the time (see Chapter 4). They use different lines of evidence wherever possible to strengthen inferences. But phylogenies represent hypothetical historical relationships; they are explanations for patterns in nature that scientists can test with further evidence. And the information used to generate phylogenies can be voluminous, investigating a diversity of genes, populations, or species (see Chapter 7). Statistical methods help scientists resolve their data and develop the best possible hypotheses.

2. Which of the following statements about molecular clocks is FALSE?

   a. Molecular clocks cannot be used to measure divergence in species separated by more than a few hundred million years.

   Correct. Molecular clocks can be used to measure divergence in species separated by hundreds of millions of years, but clocks must be calibrated because different types of DNA segments evolve at different rates. For species separated by hundreds of millions of years, slow-evolving segments of DNA will provide greater accuracy than fast-evolving segments because over time, patterns in fast-evolving segments will be more and more difficult to discern (see Chapter 7).

   b. Molecular clocks must be calibrated because different types of DNA segments evolve at different rates.

   Incorrect. Molecular clocks result because within DNA, base-pair substitutions accumulate at a roughly clocklike rate, but different types of DNA segments evolve at different rates. This type of variation is predicted by neutral theory (see Chapter 7).

c. Molecular clocks result because within DNA, base-pair substitutions accumulate at a roughly clocklike rate, although substitution rates may differ between lineages.

Incorrect. Base-pair substitutions do accumulate at a roughly clocklike rate, and these substitution rates can differ between lineages. This type of variation is predicted by neutral theory (see Chapter 7).

d. Molecular clocks can be used to predict and test the ages of unknown samples of DNA.

Incorrect. Molecular clocks can and have been used to predict and test ages of unknown samples of DNA. In fact, scientists were able to accurately predict the age of an HIV-1 sample dating back more than 40 years using a molecular clock (see Chapter 7).

e. None of the above is a false statement about molecular clocks.

Incorrect. Molecular clocks can be used to measure divergence in species separated by hundreds of millions of years. They result because base-pair substitutions accumulate at a roughly clocklike rate, but different types of DNA segments evolve at different rates. Also, substitution rates can differ between lineages. More importantly, molecular clocks can and have been used to accurately predict and test ages of unknown samples of DNA (see Chapter 7).

3. Which is NOT a benefit of living in a social group?

a. Reduced time spent watching for predators.

Incorrect, a benefit. Although it seems that reducing any time watching for predators would increase an individual's likelihood of being killed, in a group there are more eyes to watch for predators. So at any one time, some individual in the group is likely watching for predators. As a result, any single individual has to spend less time watching, leaving more time to forage (or pursue matings or snooze). Spending less time watching for predators is a benefit of living in a social group.

b. Large brain size.

Correct, NOT a benefit. Brain size may have evolved in response to selective pressures associated with living in a social group. Indeed, scientists have found that primate species that live in large groups tend to have proportionately larger neocortexes than those species that live in small groups (Figure 13.24). But large brain size is not necessarily a benefit of living in a social group.

c. Increased inclusive fitness.

Incorrect, a benefit. Inclusive fitness incorporates an individual's own reproduction as well as any increase in the reproduction of its relatives due specifically to its own actions. Living in a social group may have a cost because individuals are required to help raise offspring that are not their own, but if those offspring are relatives, the helping individuals are increasing the likelihood that some of their genes will be passed on to future generations. Helping has the added bonus of providing experience and opportunities to assess mates—both of which can raise the individual's reproductive success (see Chapter 13).

d. Smaller risk to individuals of being killed by a predator.

Incorrect, a benefit. Living in a group can reduce the risk that any particular individual within the group is killed by a predator. That reduced risk is called the dilution effect, and it can be a great benefit of living in a social group. However, group living incurs costs such as increased conspicuousness to predators, increased competition for food or mates, or decreased confidence in paternity (see Chapter 13).

e. Enhanced opportunities for learning.

Incorrect, a benefit. Learning may be a benefit of living in social groups because opportunities to observe may be enhanced. For example, juvenile New Caledonian crows stay with their parents for over a year, and that extended stay may allow them time to learn how to use tools instead of having to invent tools themselves. Group living can incur costs, however, such as increased competition for food or mates and can increase conspicuousness to predators (see Chapter 13).

Identify Key Terms

Acheulean technology	A distinctive method for making large, teardrop-shaped stone tools called hand axes. The technique was much more sophisticated than earlier tool-making styles, but changed little from 1.7 million years ago to 100,000 years ago when hand axes disappeared from the fossil record.
australopith	A paraphyletic group of hominins that lived in Africa between 4 and 2 million years ago, including members of the genus *Australopithecus* (e.g., *Australopithecus afarensis*).
bipedalism	A form of locomotion where an organism walks upright on two legs.
gene duplication	A mutation that causes a segment of DNA to be copied a second time. The mutation can affect a region inside a gene, an entire gene, and in some cases, an entire genome.
gene expression	The process by which information from a gene is transformed into a product.
gene recruitment	The co-option of a particular gene for a totally different function as a result of a mutation. The reorganization of a preexisting regulatory network can be a major evolutionary event.
hominin	Members of the "human" branch of the hominid clade, including the genus *Homo* and its close relatives, such as *Australopithecus*, but not chimpanzees and bonobos (*Pan*), gorillas (*Gorilla*), and orangutans (*Pongo*).
Neanderthal	A Eurasian lineage of hominins that lived from 300,000 to about 28,000 years ago and likely interbred with humans.
ultra-social hypothesis	A hypothesis that suggests that hominins, and humans in particular, evolved from other apes because selection favored ever-increasing social skills necessary for competing and cooperating in groups.

Interpret the Data

- If ultra-social relationships acted as selective agents on human brain development, would you predict that children, chimpanzees, and orangutans should differ in their understanding of basic math? Why or why not?

  No. Children should not do better than chimpanzees and orangutans with tests of skills like math because the ultra-social hypothesis makes no prediction about humans, math skills, and development. Humans and other apes share a common ancestor, a common ancestor that experienced selection for complex cognition, including tool use and sophisticated forms of communication and perhaps even recognition. But the ultra-social hypothesis predicts that humans are fundamentally different from other apes in social cognition. Children should do better than chimpanzees and orangutans on tests related to observation and learning because selection favored rapid development of social cognition in humans.

- What do the width of the green bars and the "whiskers" for children, chimpanzees, and orangutans indicate?

  Variation among individuals in test scores.

- If the whiskers indicate individuals with test results at the ends of the distributions of each species, how would you interpret the differences among children and chimpanzees in social skills test?

  A few individual children had test results that overlapped with a few chimpanzee test results, but the majority of children scored higher than the majority of chimpanzees.

Delve Deeper

1.

1. Which hypothesis has more support for the evolution of bipedalism: efficient feeding or heat dissipation?

Scientists have proposed two hypotheses for the evolution of walking upright: (1) that it permitted more efficient feeding, and (2) that it allowed greater heat dissipation. Individuals that had more stable footing could reach and stretch for fruits, thereby increasing foraging efficiency. Alternatively, individuals walking with an upright stride would expose less of their skin to the sun and would have access to cooler air away from the Earth's surface. Both the efficient feeding hypothesis and the heat dissipation hypothesis are still being tested, so currently neither has more support than the other. In addition, the two hypotheses might not be mutually exclusive—they might both have influenced the evolution of bipedalism.

2. Would you expect the common ancestor of chimpanzees and early hominins to have used tools? Why or why not?

Yes, the common ancestor of chimpanzees and early hominins likely used tools. Chimpanzees learn and use tools, and ample evidence exists to indicate that early hominins did so as well. In fact, different types of tools can be associated with different hominin species. The evolution of tool use may be associated with lineages that already have complex social cognition and unusually large brains. Evidence indicates that early hominins lived in social groups, like chimpanzees, and shared behavioral and physiological adaptations for living in those circumstances. Indirect evidence also indicates that even the cultural evolution of tools across chimpanzee groups is similar to the cultural evolution of tools in modern-day humans. So the most parsimonious explanation for these shared abilities in chimpanzees and early hominins is that the common ancestor of the two lineages could also use tools.

3. What evidence indicates that our species, Homo sapiens, split from *Homo heidelbergensis* and not *Homo neanderthalensis*?

Both the fossil record and DNA evidence indicate that *Homo sapiens* split *from Homo heidelbergensis* and not *Homo neanderthalensis*. Fossil evidence indicates that

Homo heidelbergensis spread from Africa across Europe about 600,000 years ago, before the evolution of *Homo sapiens* (before the oldest fossil evidence belonging to *Homo sapiens*). The oldest fossils of *Homo neanderthalensis* were found in Europe and date to about 300,000 years ago. Neanderthals represented a new lineage distinguishable from *Homo heidelbergensis*, and the species spread across the continent until relatively recently. Bones and artifacts have been discovered in a rock shelter in Jordan from Neanderthals who lived between 69,000 and 49,000 years ago and in a cave in Spain from 40,000 years ago. The oldest fossil evidence belonging to *Homo sapiens* dates back to 200,000 and 160,000 years ago from Ethiopia, indicating that *Homo sapiens* and *Homo neanderthalensis* existed at the same time.

DNA evidence provides additional support for the hypothesis that Neanderthals and humans represent two separate lineages of Homo descending from a common ancestor. Comparisons of alleles between the Neanderthal genome and the human genome points to a shared common ancestor that lived about 400,000 years ago. Additional DNA evidence indicates that Neanderthals also share a common ancestor—to the exclusion of humans. All of the humans share a recent common ancestor, which molecular clock estimates place at roughly 150,000 years ago. And variation in the DNA of living humans shows that Africans have the greatest genetic diversity, suggesting that *Homo sapiens* evolved in Africa and then expanded out to other continents.

4. Which molecule(s) related to our emotions likely evolved early in our mammalian history?

Dopamine is a neurotransmitter that arouses an animal's attention. Rats and other mammals share this neurotransmitter, although they may respond differently than humans. Oxytocin is a hormone that elicits bonding behavior in mammals, including humans. Smell usually causes the release of this hormone in other mammals, but in humans, release is related to sight, which is consistent to other shifts toward sight in the apes.

5. Why isn't the human brain a rational decision-making organ like a computer?

The human brain isn't a rational decision-making organ because the genetic architecture of the human brain did not evolve to be rational. Evolution of the human brain occurred in response to selective pressures that optimized fitness, not necessarily the ability to reason. As humans evolved, hominin group sizes increased, and our ancestors became highly social, interacting with individuals with complex behaviors. Natural selection favored structures and pathways in the brain that enhanced survival and reproductive fitness within this system. But the function of the brain is limited by what can develop structurally as well as the by the limitations of co-opted systems and pathways. Often, structures that develop that guide behavior and enhance survival in one instance may not be easily restructured simply because humans need them to be. Understanding that our brain does not function like the rational brain of a species like *Homo economicus* may help us understand the selective pressures that led to the development of behaviors, such as the strong loss-averse behavior where we will take more risks to hold on to what we have.

Test Yourself:
1. d; 2. f; 3. e; 4. d; 5. a; 6. e; 7. b; 8. e; 9. d; 10. a; 11. a

Chapter 15
EVOLUTIONARY MEDICINE

Check Your Understanding:

1. Which is NOT one of the three conditions necessary for evolution by natural selection to occur?
   a. Individuals must differ in the characteristics of a trait.
   b. The differences among individuals in a trait must be at least partially heritable.
   c. Some individuals survive and reproduce more successfully than others because of differences in a trait.
   d. The more an individual needs a trait, the more quickly it will adapt to its environment.
   e. All are necessary for natural selection to occur.

2. Which of the following is NOT an important factor influencing gene expression in humans?
   a. The external environment a gene is exposed to during development.
   b. Pleiotropy.
   c. Gene dominance.
   d. All of the above are important factors influencing gene expression.
   e. Only b and c are important factors influencing gene expression.

3. What would be the outcome of natural selection on a mutation that increases fertility early in life but increases susceptibility to cancerous growths later in life?
   a. Natural selection would favor individuals with the mutation because the fitness effects early in life would be bigger than the harm the mutation causes in old age.
   b. Natural selection would not favor individuals with the mutation because individuals susceptible to cancerous growths would die sooner and have lower lifetime reproductive success than individuals without the mutation.
   c. Natural selection would favor a balance between reproductive success and susceptibility to cancerous growths.
   d. Natural selection would remove individuals with the mutation from the population because mutations are detrimental.
   e. A single mutation cannot be both beneficial and detrimental.

Learning Objectives for Chapter 15:

   ➢ Explain why phylogenies are effective tools for understanding the origins of infectious diseases.
   ➢ Identify factors related to the evolutionary biology of pathogens that affect their ability to shift from one host to another.
   ➢ Discuss the public health issues associate with the evolution of influenza viruses.
   ➢ Distinguish between different kinds selection affecting the evolution of virulence.
   ➢ Demonstrate the adaptive significance of the *HbS* allele.
   ➢ Apply evolutionary theory to the development of vaccines.
   ➢ Describe the conditions necessary for the evolution of antibiotic resistance.
   ➢ Explain why an understanding of human genetic variation is important when considering drug treatment options.

> ➤ Apply models used to understand the rise of antibiotic-resistant bacteria to anticancer drug resistance.
> ➤ Analyze the trade-off between natural selection acting early in life versus late in life.
> ➤ Explain the hygiene hypothesis.
> ➤ Show how understanding our evolutionary history may help in the search for medically relevant genes.

Identify Key Terms

Connect the following terms with their definitions on the right:

antibiotic resistance	The process by which a species engaged in a coevolutionary relationship crosses an ecological barrier and becomes engaged in a new coevolutionary relationship.
autoimmune disease	Mutated versions of proto-oncogenes. Increased expression can lead to cancer.
evolutionary medicine	An event that occurs when new cases of an infectious disease occur at a rate higher than expected and across broad geographic regions. A disease that goes "viral"—literally.
genetic drift	The relative ability of a pathogen to cause disease.
horizontal gene transfer	Normal genes whose functions, when altered by mutation, have the potential to cause cancer.
host shifting	Any process in which genetic material is transferred to another organism without descent.
hygiene hypothesis	The integrated study of evolution and medicine to improve scientific understanding of the reasons for disease and actions that can be taken to improve health.
oncogenes	Evolution arising from random changes in the genetic composition of a population from one generation to the next.
pandemic	Diseases caused by our immune systems attacking our own bodies, such as Crohn's disease, asthma, and type 1 diabetes.
proto-oncogenes	The ability of a bacteria to survive exposure to an antibiotic.
virulence	A hypothesis that proposes that a lack of early childhood exposure to infectious agents, symbiotic microorganisms (e.g., gut flora or probiotics), and parasites increases susceptibility to allergic and autoimmune diseases.

Link Concepts

1. Fill in the following concept map with the key terms from the chapter:

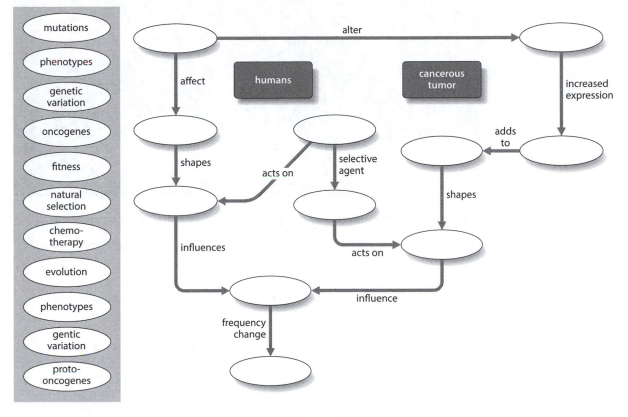

2. Develop your own concept map that explains the relationship between old age and disease. Think about concepts such as fitness, life history, natural selection, p53 tumor suppressor protein, pleiotropy, stress-fighting genes, and trade-offs.

Interpret the Data

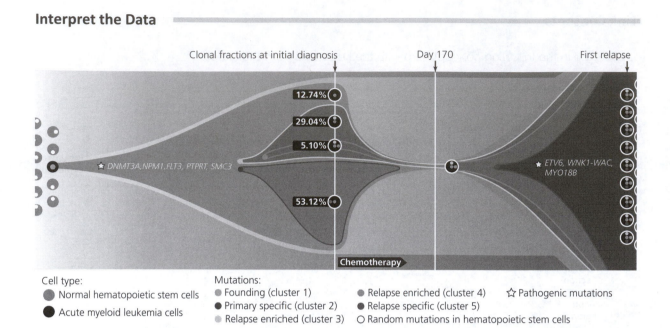

Cell type:
- Normal hematopoietic stem cells
- Acute myeloid leukemia cells

Mutations:
- Founding (cluster 1)
- Primary specific (cluster 2)
- Relapse enriched (cluster 3)
- Relapse enriched (cluster 4)
- Relapse specific (cluster 5)
- Random mutations in hematopoietic stem cells
- Pathogenic mutations

Cancer is subject to the same evolutionary pressures as other organisms. Like other organisms, three conditions are necessary for evolution by natural selection: (1) individual cancer cells must differ in their expression of a trait; (2) the differences must be at least partially heritable; and (3) some individuals must survive and reproduce more effectively than others because of these differences. Figure 15.15 illustrates the operation of these conditions in a woman who died of acute myeloid leukemia, a cancer of immune cells. The green circles on the left represent variation among hematopoietic stem cells (progenitors of the immune cells). Heritable mutations arose in one lineage, leading to faster reproduction of the cancerous cells. (The width of each wedge represents relative population size.) Within this lineage, new mutations arose leading to new lineages (the yellow and purple dots). Other mutations arose in the cluster 3 lineage, leading to yet another new lineage. Chemotherapy functioned to reduce population sizes until about Day 170, and a new mutation arose (red dot) before treatment ended. (Adapted from Ding et al. 2012.)

What happened to the cluster 3 cancer cell lineage (orange line) after chemotherapy?

What happened to the cluster 2 cancer cells after chemotherapy?

Why did the cluster 4 lineage increase so dramatically after chemotherapy?

Was the cluster 5 lineage resistant to chemotherapy?

Games and Exercises

Is It Antibiotic Resistance?

Suppose you are infected with a bacterial disease. Your sister had this same illness last week and took a full cycle of antibiotics. She quickly became better. You started taking the same antibiotic, but it had no effect. In fact, you had to return to the doctor after a week because you did not feel better. What has happened? Why did you remain sick after taking antibiotics, while your sister quickly recovered?

Three possible hypotheses explain the events:

1. You developed a tolerance for the antibiotic (i.e., you experienced a nongenetic change that made you less sensitive to the effects of the antibiotic).
2. The bacteria infecting you developed a tolerance for the antibiotic (i.e., individual bacteria experienced a nongenetic change that made them less sensitive to the effects of the antibiotic).
3. The bacteria infecting you evolved to be resistant to the antibiotic (i.e., a genetic mutation for resistance occurred in a bacterial cell, it had a reproductive advantage and increased in the population).

- Which hypothesis do you think is most likely? Why?

Now suppose that when you first visited your doctor, she told you that she is conducting research on antibiotics and asked you to be a part of the study. You agreed. As part of the study, you went to the doctor every day and let her take a new sample from your infection, which she then conducted tests on. She discovered that on the first day, your bacteria were susceptible to the antibiotic—the bacteria were killed by the antibiotic. She prescribed the antibiotic for you, which you immediately began taking. Later in the week, the bacteria from your infection were found to be resistant to the antibiotic—the bacteria were not killed by the antibiotic.

- Which of the three hypotheses does this result rule out (1, 2, or 3)? Why?

Another result from the study is that initially, all the bacteria were susceptible to the antibiotic, but by the third day, some of the bacteria were resistant to the antibiotic. With each passing day, more of the bacteria were resistant, until finally all of the bacteria were resistant.

- Does this result support either (or both) of the remaining hypotheses?

- Does it rule out either of them? Why or why not?

Finally, suppose your doctor performs a DNA analysis of the bacteria causing your infection and discovers that the resistant bacteria differ from the susceptible bacteria by a single gene: the gene that encodes the protein on the bacterial cell that is the "target" of the antibiotic (the target site is the place in the bacterial cell where the antibiotic binds and does its dirty work). The resistant bacteria have an altered form of this target site, with the result that the antibiotic is unable to bind to the target site and thus is unable kill the bacterium.

- How did the resistant bacteria come to be different genetically from their susceptible ancestors?

- Which hypothesis does this result support? Why?

- Which hypothesis (1, 2, or 3) do you think is most likely correct?

Evolving Resistance

Scientists collected data on antibiotic use and development of resistance from a community in Finland from 1978 to 1993. They looked at the annual amount of antibiotics used by members of the community and compared that with samples of bacteria from young children with middle ear infections. They examined the bacterial strains to see if they were susceptible to (killed by) or resistant to (not killed by) these antibiotics. The table below lists the year, the amount of antibiotics used, and the percentage of the bacterial strains that were resistant to the antibiotic (from 0 to 100 percent).

Year	Annual Antibiotic Usage	% Resistant Strains
1978	0.84	0
1979	0.92	2
1980	1.04	29
1981	0.98	46
1982	1.02	45
1983	1.03	58
1984	0.95	61
1985	1.12	60
1986	1.06	49
1987	1.14	59
1988	1.21	58
1989	1.28	71
1990	1.32	84
1991	1.31	79
1992	1.27	78
1993	1.28	91

Use this data to make two different graphs: (1) antibiotic usage vs. year (graph year on the x axis and annual antibiotic usage on the y axis); and (2) percent resistant strains vs. year (graph year on the x axis and percent resistant strains on the y axis).

year   year

- Did the annual antibiotic usage in this community increase, decrease, or stay the same over the course of the study?

- Did the percentage of bacterial strains resistant to these antibiotics increase, decrease, or stay the same over this same time period?

- From these data, do you think antibiotic usage was related to percentage of strains resistant to the antibiotics?

- Do you think one of these factors was the cause of the other factor? If you do, which one?

- How might you be able to tell for sure?

- In another study, also in Finland, researchers investigated the effect of greatly limiting the use of an antibiotic on a community (kids still got ear infections, but they now used a "wait and see" approach to see if the infection would clear up on its own). After antibiotic usage was greatly decreased, the percentage of bacterial strains that were resistant to this antibiotic was decreased by 50 percent (cut in half). What does this new information tell you?

This activity was developed by Dr. Kerry Bright at the University of Montana. If you use this lesson in your classroom, please send along some feedback (brightk@mso.umt.edu).

The Tangled Bank Study Guide

Intelligent Design: Curing Diseases with Darwinian Medicine

Richard Dawkins interviews Randolph Nesse about the genius of evolutionary medicine. Ness coauthored Why We Get Sick: The New Science of Darwinian Medicine with George C. Williams). Ness explains some important research on humans, including life-history trade-offs in men and women.

http://www.youtube.com/watch?v=EWldEn8zQ68

TEDx SantaCruz: Rachel Abrams—(R)evolutionary Medicine

TEDx is a program of local, self-organized events that bring people together to share a TED-like experience. Rachel Abrams shares her thoughts on how our evolutionary history shaped us and the medical issues facing our species today.

http://www.youtube.com/watch?v=vUP0yt-6ba4

How Is Darwinian Medicine Useful?

This short article explains how medical scientists are beginning to look at our bodies and our responses, including obesity, anxiety, symptoms, and disease.

http://www.ncbi.nlm.nih.gov/pmc/articles/PMC1071402/

Palo Alto Talks: Evolutionary Medicine

If you want to go deep into evolutionary medicine, these 17 videos take you into the inagural conference for scientists and entepreneurs sponsored by the Stanford University School of Medicine, Department of Neurology and Neurological Sciences. Sessions include disease, cancer, behavior, and mental disorders.

http://paloaltoinstitute.org/video

Big Food: Health, Culture, and the Evolution of Eating

This online exhibit from the Yale Peabody Museum of Natural History includes a video about our evolution and our food. You can play a game to help understand the balance of food intake and energy expenditure.

http://peabody.yale.edu/exhibits/big-food-health-culture-and-evolution-eating

Agent of Selection

Kevin Zelnio shares a poetic reminder of the influence of humans on our environment on the EvoEcoLab blog from Scientific American.

http://blogs.scientificamerican.com/evo-eco-lab/2011/07/21/agent-of-selection/

Common Misconceptions

Eugenics Is NOT about Science

Eugenics is the idea that people with "good" genes (the literal translation of "eugenics") should breed, at the expense of people with "bad" genes. Eugenics is a social philosophy; it's not a component of evolutionary theory.

The theory of evolution explains our human-ness; it does NOT guide it. In our early history, human races evolved because of the same principles that lead to speciation. People in the same geographic areas intermarried, local adaptations (like height or skin color) were favored, and lineages evolved. However, humans have never been very prone to isolating barriers, and "races" based on artificial classifications, such as height or skin color, are not well defined at all. Ironically, our different social customs often prevented intermarrying—not our evolutionary biology.

In our recent history, evolutionary theory has been co-opted by those who want to push their own frightening agendas. Concepts like racial "purity" and "good breeding stock" are not components of the theory—they were used by members of society for power and control, by those claiming to be superior in some way. But "superiority" is also in the eyes of the beholder, and science has very little to say about such ideas.

Evolutionary Theory Is Important to Practicing Physicians

Although most physicians don't use evolutionary medicine on a day-to-day basis, understanding evolution is critical to their everyday practice. For example, understanding the evolution of resistance is crucial to any physician prescribing antibiotics. Cancer is similarly subject to natural selection, and resistance to cancer drugs can be a problem for some patients. In addition, treatments based on evolutionary principles for autoimmune diseases, such as type I diabetes and asthma, may be just around the corner.

Contemplate

What would you predict about the evolution of antibiotic resistance if antibiotic drugs can pass through our bodies and into our urine?	Besides diet and exercise, where might scientists turn to combat obesity? Why?

Delve Deeper

1. Would you consider the *HbS* allele to be an adaptation to malaria? Why or why not?

2. How might repeated infections function to enable host shifting in some pathogens?

3. Does natural selection operate within a cancerous tumor? How?

4. Why do devastating diseases, such as Huntington's disease, continue to plague humans?

5. How did Ed Marcotte and his colleagues investigate gene function and its role in human diseases?

Go the Distance: Examine the Primary Literature

Gabriel G. Perron, Michael Zasloff, and Graham Bell examined the evolution of resistance to an antimicrobial peptide in *Escherichia coli* and *Pseudomonas fluorescens*. They were interested in specific types of antimicrobial peptides, those that interfere with the membrane structure of the bacterial cell. Using a selection experiment, Perron and his colleagues were able to show that both bacteria species independently evolved adaptations that conferred resistance to antimicrobial peptides.

- Why did evolutionary biologists hypothesize that the evolution of resistance to these specific types of antimicrobial peptides might be unlikely?

- Why are Perron's results important?

Perron, G. G., M. Zasloff, and G. Bell. 2006. Experimental Evolution of Resistance to an Antimicrobial Peptide. Proceedings of *the Royal Society Series B: Biological Sciences* 273 (1583): 251–56. doi:10.1098/rspb.2005.3301. http://rspb.royalsocietypublishing.org/content/273/1583/251.full.

Test Yourself

1. Which cytomegalovirus is the Ppyg cytomegalovirus (CMV) more closely related to?

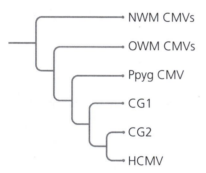

- NWM CMVs
- OWM CMVs
- Ppyg CMV
- CG1
- CG2
- HCMV

 a. Human cytomegalovirus (HCMV).
 b. New World monkey (NWM) cytomegalovirus.
 c. Old World monkey (OWM) cytomegalovirus.
 d. It is equally related to all the other cytomegaloviruses.
 e. It is not related to any other cytomegalovirus.

2. Why are scientists concerned about the H5N1 flu virus?
 a. They aren't because H5N1 is not contagious from one person to the next.
 b. Because only a few mutations may be necessary for the H5N1 virus to shift hosts.
 c. Because H5N1 is especially virulent.
 d. Because H5N1 contains only 10 genes on eight segments of RNA, and it can replicate quickly.
 e. They aren't. H5N1 only infects birds.

3. What two opposing agents of selection influence the virulence of a pathogen?
 a. Selection for high mutation rates and selection against detrimental mutations.
 b. Selection for rapid within-host replication and selection for high survival.
 c. Selection for within-host replication and selection for between-host transmission.
 d. Selection for between-host transmission and selection for between-host survival.

4. Why doesn't natural selection "weed out" the *HbS* allele from the human population if it causes sickle-cell anemia?
 a. Because the allele keeps reappearing through mutation.
 b. Because heterozygous individuals have a fitness advantage over homozygous individuals.
 c. Because natural selection can only remove recessive alleles from a population when they are rare.
 d. Because sickle-cell anemia only affects an individual's ability to survive; people with the allele can still have offspring that carry the allele.
 e. Natural selection is weeding out the allele; it just hasn't completely disappeared yet.

5. Which of these statements about vaccines is TRUE?
   a. Vaccines against flu viruses are usually ineffective because viruses have high mutation rates.
   b. Vaccines developed using cells from monkeys or chimpanzees rarely work to prevent disease in humans.
   c. Vaccines can be developed by altering the selective environment that a virus is adapted to.
   d. Vaccines rarely function to prevent disease because viruses can adapt so quickly.

6. What does the fact that the first antibiotic (Rifampin) given to JH to treat his *Staphylococcus aureus* infection failed to work indicate?
   a. That the strain of *Staphylococcus aureus* JH had become infected with may have already evolved resistance to some antibiotics.
   b. That natural selection had favored strains of Staphylococcus aureus within JH that could mutate rapidly.
   c. That JH was reinfected after treatment by a new strain that doctors failed to detect.
   d. That Rifampin is not a functional antibiotic.
   e. That the strain of Staphylococcus aureus JH had become infected with would eventually evolve resistance to every antibiotic.

7. What factors associated with the evolutionary biology of bacteria facilitate the evolution of resistance to antibiotics?
   a. High mutation rates.
   b. Horizontal gene transfer.
   c. Sub-lethal doses of antibiotics.
   d. All of the above facilitate the evolution of resistance.
   e. None of the above facilitates the evolution of resistance.

8. Why does complete color blindness affect the population of the island of Pingelap over 1500 times more than the population of the United States?
   a. Because of a population bottleneck that reduced the population to 20 individuals.
   b. Because color blindness is caused by excessive exposure to sunlight.
   c. Because so many people live in the United States, color blindness is hard to diagnose.
   d. All of the above are causes of the high rate of color blindness.
   e. None of the above is a cause of the high rate of color blindness.

9. Which statement is NOT a prediction of the hygiene hypothesis?
   a. Children who frequently play outside should develop allergic diseases less often than children who don't play outside.
   b. Children who never wash their hands or bathe will be less likely to develop immune disorders than children who wash regularly.
   c. The more infectious diseases an individual is exposed to as a child, the less likely that person will be to develop immune disorders as an adult.
   d. Extensive use of antibiotics as a child may affect how the body responds to future bacterial invasions.
   e. Pathogens that have long historical associations with humans, such as *H. pylori*, should be more positively related to immune system function than pathogens with more recent associations.

10. Why is understanding evolutionary biology important to the future of medicine?
    a. Because phylogenies can be used to predict which influenza viruses may be particularly likely to cause future epidemics.
    b. By examining analogous genes in other organisms, scientists may be able to identify the underlying genetic architecture of human disease.
    c. Evolutionary biology can guide the development of antibiotic treatments, antiviral drugs, and chemotherapy.
    d. All of the above.
    e. None of the above.

Answers for Chapter 15

Check Your Understanding

1. Which is NOT one of the three conditions necessary for evolution by natural selection to occur?
   a. Individuals must differ in the characteristics of a trait.

   Incorrect. Individual variation in the expression or characteristics of a trait is the raw material for evolution by natural selection. The ultimate source of that variation is mutations to the genetic architecture within an individual. Some mutations are detrimental, some are beneficial, and some are neutral (although additional mutations can change that). How the genotype (the genetic makeup of an individual) affects the phenotype (the observable, measurable characteristic) can be complex, but the phenotype is the manifestation of individual differences on which natural selection can act (Chapter 5). Heritable variation is integral to natural selection for most organisms because it is an effective mechanism by which beneficial mutations may be transmitted. Then natural selection acts to alter the abundances of those characteristics in the population based on the relative reproductive success of individuals possessing them (see Chapter 6).

   b. The differences among individuals in a trait must be at least partially heritable.

   Incorrect. Heritable variation is integral to natural selection for most organisms because it is an effective mechanism by which genotypic variation may be transmitted. Genotypic variation arises from mutations. Whether they be detrimental, neutral, or beneficial, mutations that affect the phenotype can affect survival and reproduction and therefore be passed to offspring or not (Chapter 5). Natural selection acts to alter the abundances of those mutations in the population based on the relative reproductive success of individuals (see Chapter 6).

   c. Some individuals survive and reproduce more successfully than others because of differences in a trait.

   Incorrect. Natural selection acts on the relative success of individuals within a population. Individuals with heritable phenotypic traits that confer some advantage or disadvantage within the population will, on average, have higher or lower survival and/or reproductive rates than other individuals who possess different versions of those traits. Natural selection acts to alter the abundances of those traits, leading to evolution of the population (see Chapter 6).

   d. The more an individual needs a trait, the more quickly it will adapt to its environment.

   Correct, NOT a condition. Evolution by natural selection is not driven by need. An individual cannot compel itself or its offspring to do better. A whale may need fins to swim, but the ancestors of whales did not necessarily need fins to swim. Fins in whales evolved over hundreds of thousands of years as mutations that affected the ancestral whale phenotype and influenced the relative survival and reproductive success of individuals as they hunted in a new, watery habitat. Certainly, the greater the influence those phenotypic changes had on the relative success of individuals, the more likely individuals in the next generation carried those mutations, and the more likely they were to be more successful in that new habitat than individuals without those mutations. However, evolution cannot identify needs, and need does not influence the evolutionary response to selection (see Chapter 2 for other misconceptions about evolution).

   e. All are necessary for natural selection to occur.

   Incorrect. For evolution by natural selection to occur, individuals must differ in the characteristics of a trait, and those differences must be heritable. Ultimately, some individuals survive and reproduce more successfully than others because of those differences (see Chapter 6). However, evolution by natural selection is not driven by need. An individual cannot compel itself or its offspring to do better.

2. Which of the following is NOT an important factor influencing gene expression in humans?

a. The external environment a gene is exposed to during development.

Incorrect. The external environment a developing individual is exposed to can indeed play an important role in gene expression. The temperature an individual develops in, the food available to it, and the infections it suffers can all influence gene expression and development. In fact, different environments can lead to very different phenotypes from a single genotype. The "environment" is not restricted to the external conditions around us, however. The environment can be anything that interacts with the promoter region of a gene in a way that influences whether that gene is expressed (see Chapter 5).

b. Pleiotropy.

Incorrect. Pleiotropy occurs when a single gene affects the expression of many different phenotypic traits; antagonistic pleiotropy is when the beneficial effects for one trait cause detrimental effects on other traits (see Chapter 8). So pleiotropic effects can be very important factors influencing gene expression in humans. For example, animals as different from us as insects use the same "genetic toolkit" to regulate development. Genes at the top of these regulatory hierarchies influence many others, and mutations in a few of these regulatory genes can produce far-reaching changes, including limb loss and the development of novel traits (see Chapter 8).

c. Gene dominance.

Incorrect. Dominant alleles are the alleles that manifest in the phenotype in a heterozygote, and they can be important factors influencing gene expression. Complete dominance occurs when the phenotype of the heterozygote is identical to the homozygote. So an individual with a completely dominant allele for smooth peas will produce smooth peas whether or not the offspring are homozygous or heterozygous for smooth peas (see Figure 5.12). However, complete dominance is rare, and expression of many human traits, such as hair color, eye color, and disease, is likely complex.

d. All of the above are important factors influencing gene expression.

Correct. The temperature an individual develops in, the food available to it, the

infections it suffers can all influence gene expression and development. In fact, different environments can lead to very different phenotypes from a single genotype (see Chapter 5). Pleiotropy occurs when a single gene affects the expression of many different phenotypic traits. Genes at the top of regulatory hierarchies influence many others, and mutations in a few of these regulatory genes can produce far-reaching changes (see Chapter 8 for an overview). Dominant alleles are the alleles that manifest in the phenotype in a heterozygote, and they can be important factors influencing gene expression. However, complete dominance is rare, and expression of many human traits, such as hair color, eye color, and disease, is likely complex.

e. Only b and c are important factors influencing gene expression.

Incorrect. Pleiotropy and dominance are clearly important factors influencing gene expression in humans, either through their effects on the expression of other genes or on their expression of a single gene. However, external environmental factors, such as the temperature an individual develops in, the food available to it, and the infections it suffers can all influence gene expression and development. In fact, different environments can lead to very different phenotypes from a single genotype, a phenomenon known as phenotypic plasticity (see Chapter 5).

3. What would you predict would be the outcome of natural selection on a mutation that increases fertility early in life but increases susceptibility to cancerous growths later in life?

a. Natural selection would favor individuals with the mutation because the fitness effects early in life would be bigger than the harm the mutation causes in old age.

Correct. Because of antagonistic pleiotropy, a mutation can result in beneficial effects for one trait, such as increased fertility, and detrimental effects on other traits, such as increased susceptibility to cancerous growths. Natural selection should favor the optimal trade-off that maximizes the number of offspring surviving to maturity over the course of an organism's entire life. So according to life-history theory, the fitness benefits of an allele that increases fertility early in life may overcome the increased probability of death from cancerous growths later in life, especially if the detrimental effects of the allele manifest after the organism has finished reproducing. Individuals with the mutation would have relatively greater reproductive success, leaving more offspring even if they might die a little sooner than individuals without the mutation (see Chapter 9).

b. Natural selection would not favor individuals with the mutation because individuals susceptible to cancerous growths would die sooner and have lower lifetime reproductive success than individuals without the mutation.
Incorrect. Many genes are pleiotropic—they affect the expression of many different phenotypic traits, so a mutation to a pleiotropic gene can have beneficial effects for one trait and detrimental effects on other traits. Because of antagonistic pleiotropy, a mutation can result in beneficial effects for one trait, such as increased fertility, and detrimental effects on other traits, such as increased susceptibility to cancerous growths. Natural selection should favor the optimal trade-off that maximizes the number of offspring surviving to maturity over the course of an organism's entire life. So according to life-history theory, even though an allele may cause an increased probability of death from cancerous growths later in life, the fitness benefits of increased fertility can be enough to maximize lifetime reproductive success, especially if the extrinsic mortality rate is high. Individuals with the mutation would have relatively greater reproductive success, leaving more offspring even if they might die a little sooner than individuals without the mutation (see Chapter 9).

c. Natural selection would favor a balance between reproductive success and susceptibility to cancerous growths.
Incorrect. Pleiotropy occurs when a single gene affects the expression of many different phenotypic traits. A mutation to a pleiotropic gene can have beneficial effects for one trait and detrimental effects on other traits—a phenomenon known as antagonistic pleiotropy. So a mutation to a single gene can result in increased fertility and increased susceptibility to cancerous growths. Natural selection should favor the optimal trade-off that maximizes the number of offspring surviving to maturity over the course of an organism's entire life. According to life-history theory, the fitness benefits of an allele that

increases fertility early in life may overcome the increased probability of death from cancerous growths later in life, especially if the detrimental effects of the allele manifest after the organism has finished reproducing. Individuals with the mutation would have relatively greater reproductive success, leaving more offspring even if they might die a little sooner than individuals without the mutation (see Chapter 9).

d. Natural selection would remove individuals with the mutation from the population because mutations are detrimental.
Incorrect. Mutations can be beneficial, neutral, or detrimental (see Chapter 5). More importantly, many genes are pleiotropic—they affect the expression of many different phenotypic traits, so a mutation to a pleiotropic gene can have beneficial effects for one trait and detrimental effects on other traits. Because of antagonistic pleiotropy, a mutation can result in beneficial effects for one trait, such as increased fertility, and detrimental effects on other traits, such as increased susceptibility to cancerous growths. Natural selection should favor the optimal trade-off that maximizes the number of offspring surviving to maturity over the course of an organism's entire life. According to life-history theory, the fitness benefits of an allele that increases fertility early in life may overcome the increased probability of death from cancerous growths later in life, especially if the detrimental effects of the allele manifest after the organism has finished reproducing. Individuals with the mutation would have relatively greater reproductive success, leaving more offspring even if they might die a little sooner than individuals without the mutation (see Chapter 9).

e. A single mutation cannot be both beneficial and detrimental.
Incorrect. Pleiotropy occurs when a single gene affects the expression of many different phenotypic traits, so a single mutation to a pleiotropic gene can have beneficial effects for one trait and detrimental effects on other traits—a phenomenon known as antagonistic pleiotropy. Because of antagonistic pleiotropy, a mutation can result in increased fertility and increased susceptibility to cancerous growths. Natural selection should favor the optimal trade-off that maximizes the number of offspring surviving to maturity over the course

of an organism's entire life. According to life-history theory, the fitness benefits of an allele that increases fertility early in life may overcome the increased probability of death from cancerous growths later in life, especially if the detrimental effects of the allele manifest after the organism has finished reproducing. Individuals with the mutation would have relatively greater reproductive success, leaving more offspring even if they might die a little sooner than individuals without the mutation (see Chapter 9).

Identify Key Terms

antibiotic resistance	The ability of bacteria to survive exposure to an antibiotic.
autoimmune disease	Diseases caused by our immune systems attacking our own bodies, such as Crohn's disease, asthma, and type 1 diabetes.
evolutionary medicine	The integrated study of evolution and medicine to improve scientific understanding of the reasons for disease and actions that can be taken to improve health.
genetic drift	Evolution arising from random changes in the genetic composition of a population from one generation to the next.
horizontal gene transfer	Any process in which genetic material is transferred to another organism without descent.
host shifting	The process by which a species engaged in a coevolutionary relationship crosses an ecological barrier and becomes engaged in a new coevolutionary relationship.
Hygiene Hypothesis	A hypothesis that proposes that a lack of early childhood exposure to infectious agents, symbiotic microorganisms (e.g., gut flora or probiotics), and parasites increases susceptibility to allergic and autoimmune diseases.
oncogenes	Mutated versions of proto-oncogenes. Increased expression can lead to cancer.
pandemic	An event that occurs when new cases of an infectious disease occur at a rate higher than expected and across broad geographic regions. A disease that goes "viral"—literally.
proto-oncogenes	Normal genes whose functions, when altered by mutation, have the potential to cause cancer.
virulence	The relative ability of a pathogen to cause disease.

Link Concepts

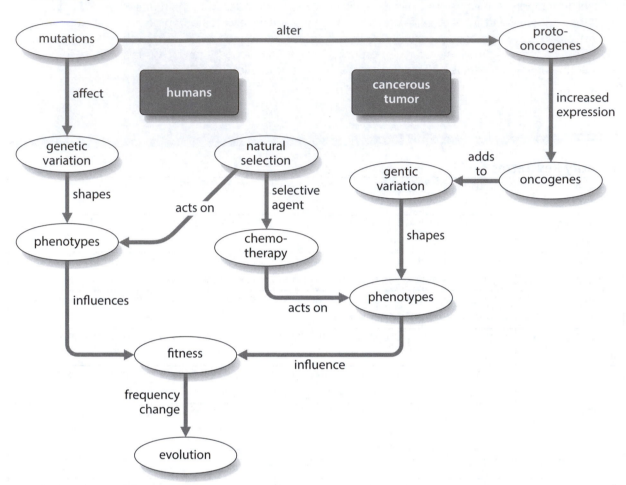

Interpret the Data

- What happened to the cluster 4 cancer cell lineage (orange line) after chemotherapy?
  The lineage survived chemotherapy.
- What happened to the cluster 2 cancer cells after chemotherapy?
  The lineage went extinct.
- Why did the cluster 3 and cluster 4 lineages increase so dramatically after chemotherapy?
  Because they had mutations that allowed them to be resistant to chemotherapy, they were able to survive and reproduce at a high rate, especially since other cancer lineages were eliminated.
- Was the cluster 5 lineage resistant to chemotherapy?
  Yes. The lineage arose from mutations to the cluster 4 lineage, a lineage that survived chemotherapy because of a resistance trait. The cluster 5 lineage came to dominate the population, causing the first cancer relapse.

Delve Deeper

1. Would you consider the *HbS* allele to be an adaptation to malaria? Why or why not?
   Yes, the *HbS* allele could be considered an adaptation to malaria. Adaptations are heritable traits that allow an individual to outcompete other individuals that lack that trait. Although individuals carrying two copies of the *HbS* allele suffer from sickle-cell anemia, individuals carrying a single *HbS* allele are only one-tenth as likely to get

severe malaria, a major fitness advantage. This heterozygote advantage is apparent in populations where malaria is prevalent—the *HbS* allele occurs at much higher frequencies than in areas where malaria is less common. In addition, the high level of linkage disequilibrium in the DNA surrounding the HbS allele indicates strong recent natural selection. So although the *HbS* allele can reduce fitness in homozygotes, clearly natural selection is acting to maintain the HbS allele in populations regularly exposed to malaria.

2. How might repeated infections function to enable host shifting in some pathogens?
A pathogen infecting a new host faces strong selective pressure. If individuals survive, genetic variation in the new population will be relatively low. Beneficial mutations within this new population may not arise quickly enough to overcome the new host's immune response. However, multiple infections may bring additional individuals with mutations beneficial for surviving on the new host, and horizontal transfer may bring new combinations of alleles together. Genetic variation in the pathogen population on the new host would increase with each new infection, and this relatively rapid increase in genetic variation may lead to rapid increases in fitness in the new population. As a result, repeated infections might provide the essential genetic variation to the pathogen population to enable host shifting.

3. Does natural selection operate within a cancerous tumor?
Yes. For an evolutionary response to natural selection, three conditions are necessary: traits must vary among individuals, that variation must have a heritable component, and individuals possessing certain traits must fare better than individuals who possess other traits. This model can be applied to the evolution of a cancerous tumor and the development of resistance to chemotherapy. Once a proto-oncogene

mutates into an oncogene, a cell can grow and divide faster than its neighbors. The mutation gives the cell an immediate fitness gain, driving the evolution of increased cell division and metastasis at the expense of the host.

Resistance to chemotherapy and drugs can evolve in a similar fashion. Some cell lineages may be less susceptible to drugs than others, or mutations may arise that confer resistance. Once this kind of heritable variation arises, natural selection will strongly favor individual cells that can resist drugs, leading to faster growth than susceptible cancer cells. Ultimately, these resistant cells will dominate the tumorous growth. Paradoxically, the evolution of these traits not only leads to the death of the host but to all of the cancer cells as well.

4. Why do devastating diseases, such as Huntington's disease, continue to plague humans?
Not all diseases affect fitness as soon as an organism is born. In fact, some diseases may only become issues because we now have longer lifespans than in the past. A disease such as Huntington's usually doesn't show signs until individuals are past breeding age. Individual with Huntington's can pass on the alleles to their offspring before they even know they carry the disorder.

Many diseases may involve a similar trade-off. Natural selection may favor some traits early in life, but those traits may be detrimental later in life. For example, production of the p53 tumor suppressor protein may prevent cancer in young individuals, allowing them to survive and reproduce. However, the p53 tumor suppressor protein may negatively interact with surrounding tissues. As we age, these negative interactions accumulate, and once we've passed breeding age, natural selection will have little effect.

5. How did Ed Marcotte and his colleagues investigate gene function and its role in human diseases?

Marcotte and his colleagues used our shared history with other organisms to look for gene networks that were relatively similar. They identified several genes in yeast that were involved with cell-wall repair, and those genes matched genes in vertebrates involved with blood-vessel repair. Then they examined the network of genes related to cell-wall repair in yeast and used that network to identify additional genes in vertebrates with unknown functions. They experimentally manipulated the genes with unknown functions in frog embryos and found that these unknown genes affected blood-vessel repair, just as their known counterparts had. Marcotte and his colleagues used those results to predict how the genes might function in humans. They also used this understanding to experiment with a drug that might be useful in the treatment of cancerous tumors.

Test Yourself:
1. a; 2. b; 3. c; 4. b; 5. c; 6. a; 7. d; 8. a; 9. b; 10. d